OLED Microdisplays

Series Editor
Robert Baptist

OLED Microdisplays

Technology and Applications

Edited by
François Templier

iSTE

WILEY

First published 2014 in Great Britain and the United States by ISTE Ltd and John Wiley & Sons, Inc.

ISTE Ltd
27-37 St George's Road
London SW19 4EU
UK

www.iste.co.uk

John Wiley & Sons, Inc.
111 River Street
Hoboken, NJ 07030
USA

www.wiley.com

Library of Congress Control Number: 2014941990

British Library Cataloguing-in-Publication Data
A CIP record for this book is available from the British Library
ISBN 978-1-84821-575-7

MIX
Paper from responsible sources
FSC® C013604
FSC www.fsc.org

Printed and bound in Great Britain by CPI Group (UK) Ltd., Croydon, Surrey CR0 4YY

Contents

Introduction

I.1. Revolution(s)

During the last decade, a revolution has happened in the display industry (Figure I.1). In fact, not only one, but *two* revolutions. The first revolution is the fade-out of the cathode ray tube (CRT). Like the giant meteor hitting planet earth 65 million years ago led to eradication of dinosaurs, the giant wave of flat-panel displays pushed out irremediably the 100-year old CRT.

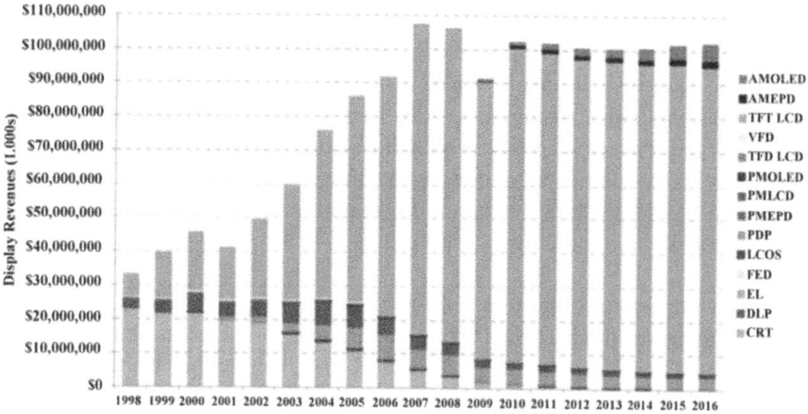

Figure I.1. *Display market 1990–2012 [NTA 14].*

The other revolution is that in the meantime, the total display market rose from a cushy ~30-billion to a tremendous 100-billion dollar market. This is illustrated by

---

Introduction written by François TEMPLIER.

the explosion of display application we have seen around us, at home (how many TV screens in the house in addition to the former single big TV in the living room?), in the car, or simply with all our digital wearable companions. During this second revolution; a very particular type of display has discreetly emerged and took a piece of the cake: microdisplays. A small piece of the market (so far), a small size, but definitely present.

I.2. Definition of microdisplays

Microdisplays can be defined as having a diagonal of around 1 inch or less. As a consequence, this means that they require an additional optical magnification system to be seen by the human eye. The system would depend on the use. In the case of near-to-eye application, such as head-mounted display (HMD) and head-up display (HUD), the system projects a virtual image, by collimating the light from the microdisplay and projecting the image near infinity. For projector applications, the collimated optics role is to make an enlarged, real image of the microdisplay on a physical diffusing screen.

One objective of microdisplays is to deliver the same image quality as conventional, larger-size displays. One of the main attributes to image quality is the *definition*, namely the total number of pixels constituting the image. If we consider the 1,080 p high-definition standard (1,080 lines and 1,920 columns), the color pixel-pitch will be around 460 μm in the case of a 40 in. diagonal TV, and around 58 μm in the case of 5 in display from a high-end smartphone. In the latter case, considering a quad color arrangement, this corresponds to an elemental pixel pitch of around 29 μm. This pixel pitch is today the smallest that can be obtained with the conventional active-matrix techniques using thin-film transistor (TFT) technology on large-area glasses, where the minimum feature size provided by the patterning process is generally around 2 μm.

To achieve the same 1,080 p resolution on a microdisplay of say 0.5 in, the pixel pitch needs to be an order of less magnitude, namely much less than 10 μm. The main consequence of such small pixel-size is that they cannot be fabricated with conventional TFT active-matrix technologies. Actually, active matrix for microdisplays requires complementary metal-oxide-semiconductor (CMOS) integrated circuit (IC) technologies built on monocrystalline silicon, which provides the smallest feature size, now down to a few nanometer for advanced circuits). Only with such technologies, pixel pitches of a few microns can be obtained. Besides the difference in active matrix technology, there is therefore also a strong difference in the manufacturing model (and cost model) between microdisplays and conventional, medium/large-size displays. Conventional displays are fabricated on large area glass (up to 3 × 3 m) and feature pixel size of typically 30–300 μm (Figure I.2: (a)

2.2 × 2.5 m glass substrate with six TFT active matrices 52 in. TVs. Inset is a magnified view of a pixel. (b) (Courtesy of Ghent University) 200 mm silicon wafer with CMOS circuit consisting of active matrices for LCOS microdisplays. Inset is a magnified view of a pixel.). Microdisplays are fabricated on silicon wafers (200 or 300 mm in diameter) and have pixel size of typically 5–10 µm (Figure I.2).

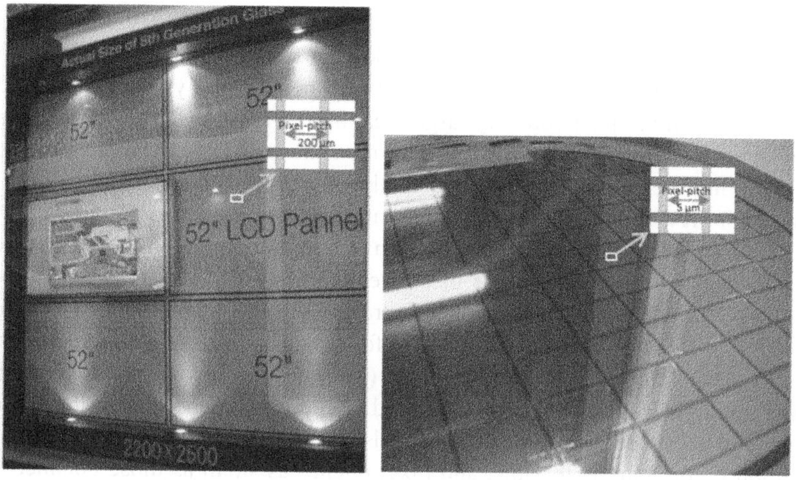

a) b)

Figure I.2. *a) 2.2 × 2.5 m glass substrate with six TFT active matrices 52 in. TVs. Inset is a magnified view of a pixel. b) (by courtesy of Ghent University) 200 mm silicon wafer with CMOS circuit consisting of active matrices for LCOS microdisplays. Inset is a magnified view of a pixel*

Table I.1 summarizes main differences between standard-type displays and microdisplays.

Type of Display	Typical size (cm)	Viewing mode	Pixel pitch (µm)	Active-matrix technology	Substrate size
Standard	5 to 200	Direct view	40 to 300	TFT on glass	Up to 3 × 3 m
Microdisplay	0.7 to 2	Magnified image	4 to 20	CMOS	Diameter of 200 or 300 mm

Table I.1. *Main differences between standard-type and microdisplays*

In summary, a microdisplay can be defined as a display needing magnifying optics to be seen, and having an IC chip as active matrix.

There might be an exception to this definition. Indeed some transmissive liquid-crystal display (LCD) microdisplays used for many years in projectors do have a small size (~0.5 to 2 in) but do not have an IC active matrix. They are fabricated using the so-called high-temperature polysilicon (HTPS) technology, which consists of making active matrix with TFTs [MOR 86]. It uses TFTs, however the approach is very different to the classical a-Si or low-temperature polysilicon (LTPS) TFT processes, which are made on very large-area, inexpensive borosilicate glass plates, and where the minimum feature size is ~2 μm or more. HTPS, as its name suggests, requires a high-temperature process: actually, it consists of depositing amorphous silicon at a few hundred degrees, and later in the process to convert it into polysilicon by performing an annealing step at a high temperature: 1000°C. Such a high temperature prohibits the use of conventional borosilicate glass, but imposes highly stable and transparent quartz substrates. These substrates are much more expensive, and therefore are relevant only for smaller sizes, actually quartz wafers with the same size as silicon counterparts (200 or 300 mm in diameter) used for CMOS circuits. Having the same size and shape as silicon wafers, these quartz wafers give access to IC manufacturing tools and design rules, therefore very high-resolution lithography is feasible, and pixel pitches as small as those obtained with IC active matrices can be made, namely ~10 μm or less. In the end, HTPS is indeed also microdisplays, since its active matrix is definitely made using IC technologies, though the substrate is not silicon, but quartz.

I.3. A brief history and overview of microdisplays

The need for microdisplays initially came from a need for bigger displays. In fact, at the time when CRT was the only display technology, it was very difficult to increase the image size and for a long time CRT TVs for household receivers did not exceed 25–30 in. Therefore, TV manufacturers were looking for solutions to increase image size, and application of projection technology was quite straightforward. CRT projection provided larger images but with a main drawback: a loss in luminance. Then came the need for the so-called "spatial light modulators" (SLM), a two-dimensional (2D) optical modulator, which would impress 2D modulation onto an optical beam. Modulation could be applied on phase, amplitude, or other. In the case of amplitude modulation, such system would perfectly fit for projecting an image.

Then the question was how to realize SLMs with satisfying performance, sufficiently thin structure, operating at acceptable supply voltages or power, etc.

I.3.1. *Liquid crystal on silicon displays*

The rise of liquid crystal made it possible. As early as the late 1970s, the first active-matrix LCD-based SLM were shown [LIP 77] and later a silicon chip was used for the active matrix: the liquid crystal on silicon (LCOS) was born. Since the silicon substrate is not optically transparent, LCOS microdisplays consisted of *reflective*-type LCDs opposite to the classical *transmissive* counterparts fabricated with glass substrates.

I.3.1.1. *Fabrication of LCOS display*

LCOS display is based on CMOS silicon active matrix, coupled with transparent cover glass and filled with liquid crystal (Figure I.3). Pixels are defined by metal electrodes on top of the CMOS, which are reflective, and covered by an alignment layer. Cover glass is coated with the type of transparent electrode (made of indium tin oxide (ITO)) and alignment layer. For such displays, a liquid-crystal can be of the reflective twisted nematic type, as it is for most commercial applications, or vertically aligned, which appeared more recently, and present advantages such as fast response, high-contrast with moderate voltage and better cell-gap tolerance [GAN 00]. Polarizer/analyzer in LCOS is obtained with a polarizer beam splitter (PBS), a key component of the optical setup, which is described later.

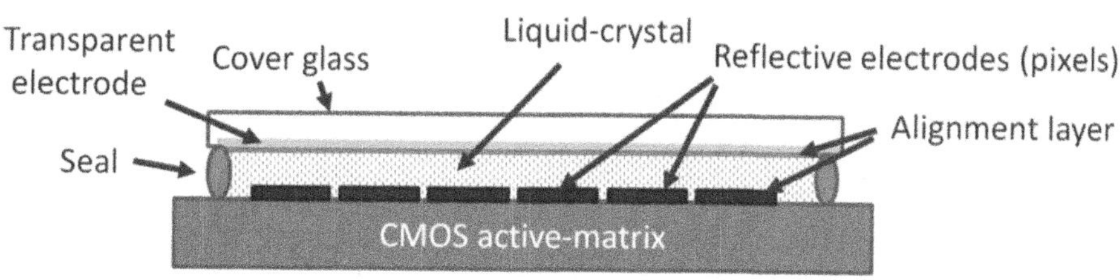

Figure I.3. *Structure of LCOS display. Only the last level of the CMOS is shown, i.e. the reflective electrodes*

Active matrix is fabricated with CMOS process. After the core MOS process, multilayer interconnects are made to build the columns, lines, buses, and the last level is the reflective electrode (Figure I.5). A pixel electrode is generally made with aluminum, a natural candidate providing very high reflection coefficient (above 90%). The top surface should be as flat as possible to help liquid-crystal operation, which is possible due to planarization techniques used now in standard interconnects processes. One advantage of reflective LCD with CMOS circuit is that a high aperture ratio (AR) can be obtained, in comparison with transmissive counterparts. In the example below, the pixel pitch is 17.6 μm and pixel space is only 0.6 μm. Even when including the small loss due to the via inside the electrode (~0.5 μm size), the dimensional AR is 93.2%. When including a reflection coefficient or 90% for Al, this gives an effective AR of as high of 83.6% [DE 02].

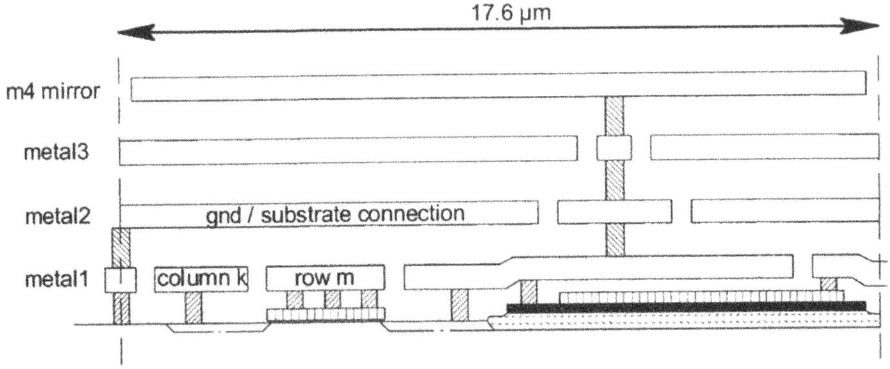

Figure I.4. *Cross section of a pixel from CMOS active matrix (from [DE 02]). The last level represents a pixel electrode (named "m4 mirror" on Figure)*

Due to high-performance of CMOS devices, gate and row drivers can be integrated on the chip, reducing cost and simplifying connection (Figure I.5). Actually, using present connection techniques, the minimum achievable connection pitch is ~50 μm; therefore it would not be possible to connect all individual lines and columns of most microdisplays.

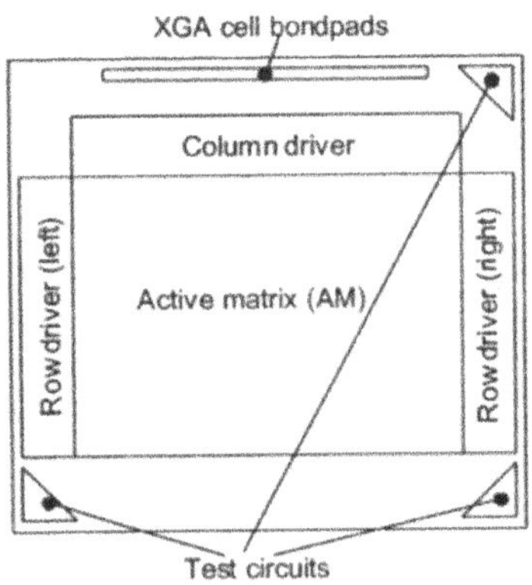

Figure I.5. *Block-circuit of active matrix for LCOS (from [DE 02])*

The cell-assembly process, which consists of coupling active matrix with transparent electrode, followed by liquid-crystal filling, can be made at wafer-scale level, or at display level. In this latter case, singulation of active matrices is made by dicing after CMOS processing (Figure I.6). Spacers might be added between the two planes to ensure perfect cell-gap control to obtain exactly the same polarization retardation/image homogeneity across the display.

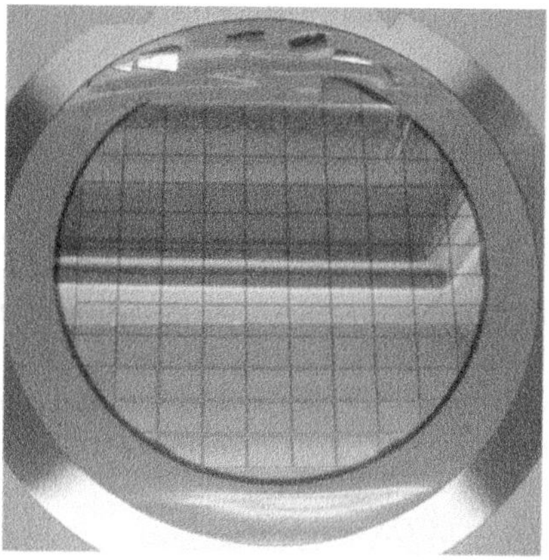

Figure I.6. *Singulation of CMOS active matrix for LCOS. Reprinted by courtesy of Ghent University*

The last step of LCOS fabrication is connecting the electronics board to the display, which is made by conventional flex-foil and anisotropic conducting film (ACF) bonding (Figure I.7).

Figure I.7. *Connection of 0.7" WXGA LCOS display. Reprinted by courtesy of Ghent University*

Depending on application, the light sources, color management and optical system will differ.

I.3.1.2. *Light sources and color management for LCOS*

Source and color options for LCOS: color can be made simply as on conventional larger size LCD displays, by using one white backlight and color filters, one color pixel requiring three (or four) RGB sub-pixels. For this reason, in this case, resolution can be limited, despite a quite simple optical system. To increase resolution while maintaining use of a single display and compactness, field-sequential color (FSC) can be used: three separate RGB LEDs (or lasers) are sequentially switched on, and respective RGB images are displayed simultaneously on the LCD. In that case, one pixel of the display is a color pixel. This requires faster electronics and fast-switching liquid crystals since one color frame (20 ms) needs three subframes (6.6 ms). The other solution is to use three light sources and three separate microdisplays, which allows full resolution without response restrictions. Actually in that case, one single white source can be used and split into three RGB beams by using dichroic filters (to separate colors).

I.3.1.3. *System integration of LCOS*

When compactness is important, one LCOS with color filters will be used, combined with a compact white backlight, or one LCOS in FCS mode. Single LCOS option is the best choice in the case of near-to-eye applications, when small size is more important than high resolution.

When a larger system is possible, and/or more power is required, one white source will be used and split in three RGB colors, combined with three displays: this is the best choice for LCOS used in projectors.

Color separation is obtained with dichroic filters, which are combined to a polarizer beam splitter (PBS) to generate linear polarization in orientation fitted for each LCOS (Figure I.8).

First-generation LCOS displays suffered from light leakage. In fact, exposure of silicon to light gives rise to carrier generation (*photoconduction*), implying that when silicon MOS transistors are under such conditions, leakage current will appear in the off-regime. If the switch transistors leaks, the information stored in the pixel is lost. Therefore, initially LCOS use was limited to application such as view finders where light levels are limited in comparison with projection. Progressively, solutions consisting of adding light-protection layers (light-shields) into the IC process were implemented, and LCOS could be applied to projection systems [MEL 98]. JVC company in Japan, has started shipping high-end professional front projectors with this technology in 1998.

Since then, LCOS microdisplays have spread and are present in near-to-eye systems as well as in small to medium-size projectors. One trend nowadays is to develop "Jumbo" LED projectors, using three power LEDS and three LCOS microdisplays, with an XCube, and providing as much as 100 lumen.

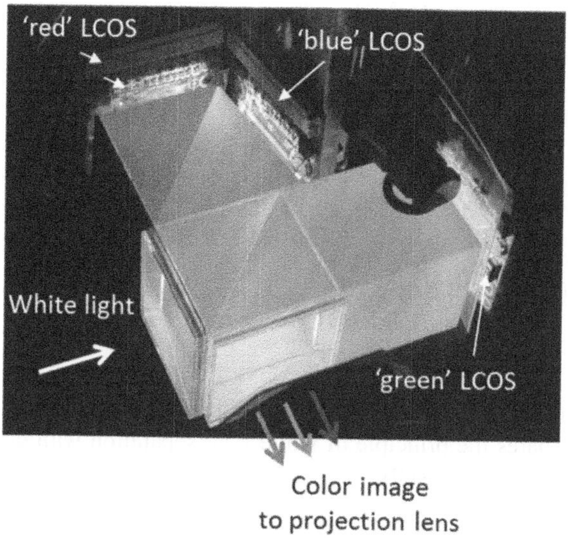

Figure I.8. *Optical setup involving a white source with three LCOS displays. Color separation and polarization are managed with dichroic filters and PBS. Reprinted by courtesy of Ghent University*

I.3.2. *Transmissive LCD microdisplays*

To fabricate microdisplay for projection, other actors have chosen the above mentioned HTPS technology, which allows fabricating high-resolution transmissive LCDs due to the transparency of the quartz substrate combined with the high precision of the IC technology for Polysilicon thin-film transistors. Such a transmissive microdisplay is more favorable for a compact optical system compared to LCOS, however obtaining high pixel aperture ratio with small pixel pitch remains a strong challenge.

As mentioned in section I.2, Seiko Epson has been proposing HTPS technology to fabricate polysilicon TFTs on a quartz substrate (Figure I.9) and has been selling transmissive LCDs for projectors with such matrices for many years [TAN 07].

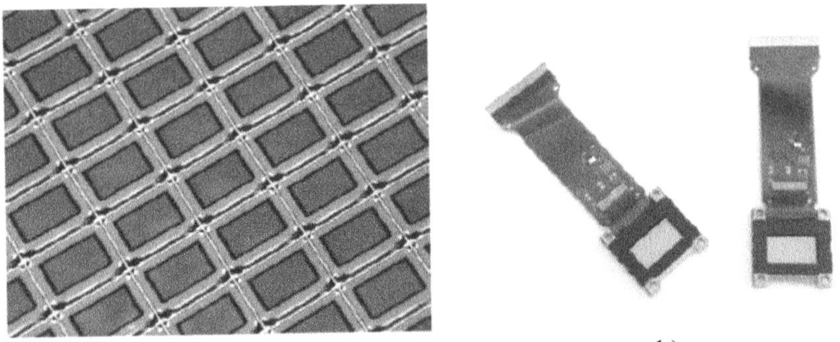

a) b)

Figure I.9. *a) HTPS active matrices on 200 mm quartz wafer and b) two finished 1920×1080 p LCD chips [PUT]*

Optical system for such displays (Figure I.10) has some differences with those for LCOS, but shares the principle of white light separation with dichroic filters. A prism recombines the three RGB images.

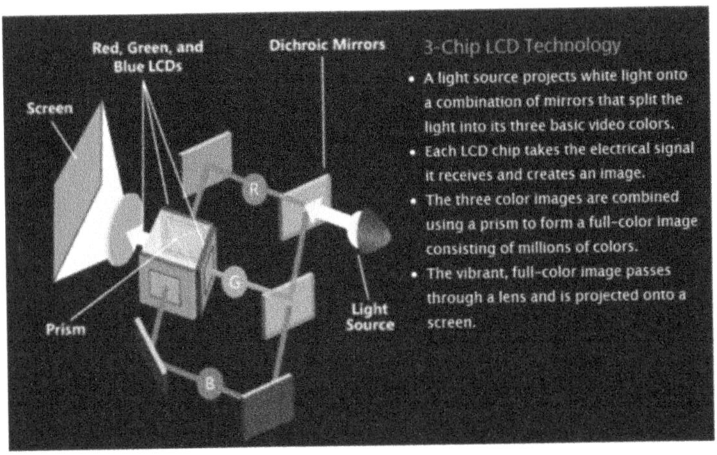

Figure I.10. *Optical setup for three transmissive LCD projector from [PUT]*

This microdisplay technology has been continuously improved for the last 20 years.

Such 3LCD systems support a wide range of video and computer resolutions. The first 3LCD video projector could display 320 × 220 pixels, while the first LCD data projector offered 640 × 480 pixels (VGA standard). Over the years, HTPS resolution has kept pace with market demand. VGA was followed by SVGA (800 × 600) in the mid-1990s, then by XGA (1204 × 768) and SXGA (1280 × 1024). While VGA and SVGA have largely gone by the wayside, the XGA standard is still popular for a wide range of business and education projectors.

The digital TV transition sparked interest in widescreen, high-definition video formats when it started in 1998. At the same time, computer manufacturers were expanding their displays to create a larger desktop workspace, just after we could see HD (1280 × 720) and SXGA+ (1400 × 1050) resolutions in projectors. Specialized large-venue projectors took those numbers even higher, achieving UXGA (1600 × 1200) pixels of picture detail (Figure I.11).

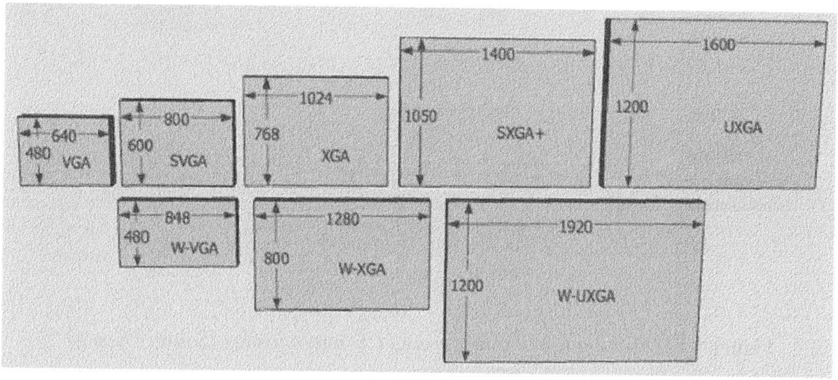

Figure I.11. *Evolution of HTPS panel size and resolution over the years. LCD chip resolutions have almost quadrupled over the 15 years. [PUT]*

Today, widescreen projection is commonplace. The 1280 × 720 format is still available for HD video projection, but it is rapidly being replaced by 1920 × 1080. And widescreen computer projection has standardized on two formats – 1280 × 800 (wide XGA) and 1920 × 1200 (wide UXGA).

Surprisingly, in 2010 Epson announced that they were fabricating HTPS-based LCDs using *reflective* mode, instead of transmissive, claiming that the open aperture was improved. Also, such a structure reduces light leakage. Compared to conventional LCOS made on CMOS circuits, HTPS might benefit from a simpler process, due to the much lower mask count of the TFT process compared to CMOS.

Another company has made transmissive LCD microdisplays for many years. Kopin, a US company, proposed an original approach: fabricate a CMOS active matrix from a silicon-on-insulator (SOI) substrate. SOI consists of a bulk silicon substrate with a thin layer of single crystalline silicon on top and separated from the substrate by a silicon dioxide layer. The SOI wafer is used because the buried oxide layer can act as an etch-stop during a subsequent back-etch step. The circuit is fabricated, and subsequently the thin Si circuit layer is lifted-off the substrate, and transferred onto a glass wafer, bonded with adhesive (Figure I.12). Using this process, a transparent active-matrix is finally made. Being fabricated with the CMOS process, very fine patterns can be made and therefore small pixel-pitch can be achieved (in fact, comparable to those obtained with LCOS).

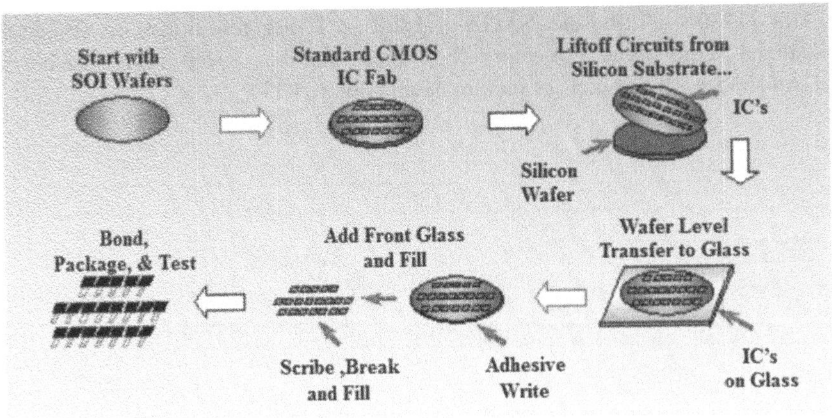

Figure I.12. *Fabrication of transparent LCD microdisplay (Source: Kopin)*

The subsequent LCD process is more conventional and similar to other LCD microdisplays. The first transmissive LCD microdisplays products were introduced in 1997. Kopin proposes these displays broad range of resolutions and size, ranging from 0.16" to 0.97", for near-to-eye applications. From the beginning, more than 30 million of these transparent LCD microdisplays have been sold.

I.3.3. *MEMS-based microdisplays*

In the same period, other kinds of IC-based SLM have been developed: micro-optical-electromechanical systems (MOEMS), in particular the digital micromirror device (DMD) developed by Texas Instruments. Such a microdisplay is fabricated on a single chip, starting with the CMOS active matrix and followed by fabrication of the electro-optical function, above-IC [HOR 89].

Actually, when it was invented in the 1970s, DMD stood for "deformable mirror device": a baseline analog device made of an array of mobile mirrors on CMOS circuit, designed to be used as analog amplitude and/or phase modulators.

Progressively, the problems encountered on such analog versions (mainly: limited dynamic range) pushed the development of digital version, based on simple two-state mirror position, which rose in the late 1980s. What then became a *digital* mirror device was continuously improved. It culminated in the launch, in 1996, of the first projector under the name "Digital Light Processing™" (DLP).

I.3.3.1. *Principle*

The principle is that a two-state mirror is present for each pixel, the on-state being a reflective state, the OFF-state on-reflective state. Actually a "flat state" is also possible. Combined with a light-source, the system can be used as a light-amplitude modulator: memory cells on the DMD loaded with a logical "1" will position the mirror, through electrostatic forces, such that the incident light is reflected into the aperture of the projection lens using dark field projection techniques as shown in Figure I.13 [DOH 98]. Memory cells loaded with a logical "0" are positioned such that the incident light is directed away from the projection lens and into a light sink. The "on" state light is optically recombined with the other pixel components and projected onto the viewing screen.

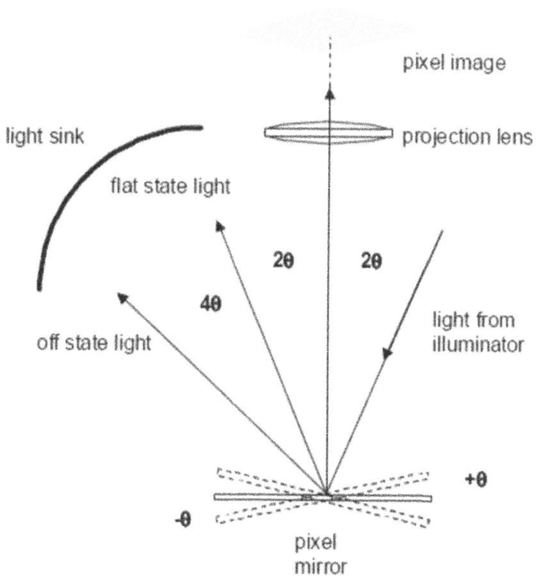

Figure I.13. *Optical setup showing elemental digital mirror device operation to be used as light-amplitude modulator. Reprinted with courtesy of the Society for Information Display*

Mirror rotation is controlled by electrostatic forces. The structure, as shown in Figure I.14, involves support structure, drive electrodes and a rotation mechanism

[SCO 03]. Such structures are fabricated above the CMOS integrated circuit, using conventional MEMS-type processes, such as 3D lithography, requiring the use of sacrificial layers and etch-stop techniques.

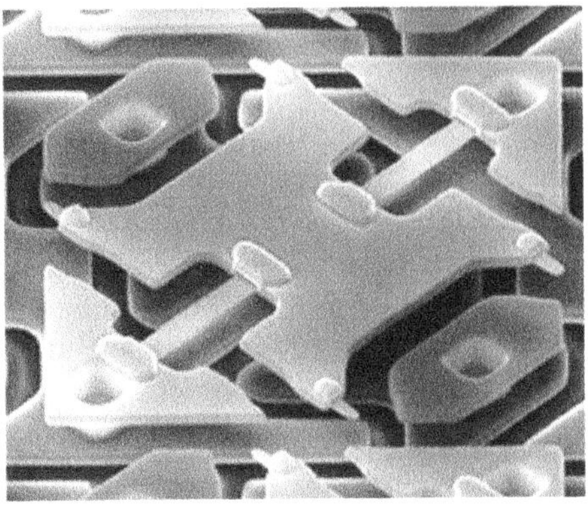

Figure I.14. *Scanning electron micrograph of DMD pixel (mirror removed). Reprinted with courtesy of the Society for Information Display*

I.3.3.2. *Grayscale management*

With such a bi-state device, grayscale is achieved by pulse width modulation (PWM) of the on-state over the operating refresh time [DOH 98], as shown in Figure I.15. The internal DLP digital (amplitude modulated) video signal is converted to this PWM. Graylevels are a function of bit number of the video data.

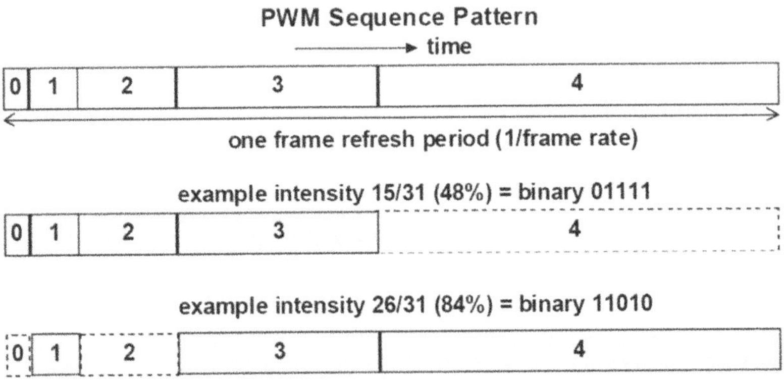

Figure I.15. *Binary PWM sequence pattern with two examples of how intensity values are generated with the sequence. For simplicity, only five-bit video is shown here (DLP1). Reprinted with courtesy of the Society for Information Display*

This pattern continues for all bits of the given pixel. The human visual system effectively integrates the pulsed light to form the perception of desired intensity. The grayscale perceived is proportional to the percentage of time the mirror is "on" during the refresh time.

The display is a full array of DMD mirrors (Figure I.16).

Figure I.16. *DMD device (Source: Texas Instruments)*

I.3.3.3. *Light source and color management*

One advantage of DLP technology is that it can use any type of light source. Historically, the main light source used on DLP display systems has been a replaceable high-pressure xenon arc lamp unit (containing a quartz arc tube, reflector, electrical connections and sometimes a quartz/glass shield), whereas when compactness is required (such as for most medium to small DLP projector) high-power LEDs or lasers are used as a source of illumination.

Due to a high switching speed, DLP is well adapted to field-sequential color. Color wheel system has been used very widely (Figure I.17) for projectors. For more compact projectors, separate RGB LEDs are preferred. Also, for high-brightness applications, three separated DLPs are used with three separate color channels.

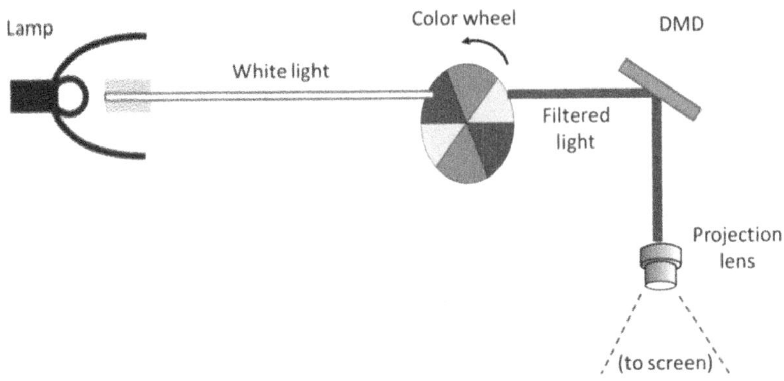

Figure I.17. *Projection system with single DMD using color wheel for field sequential color mode*

I.3.3.4. *Performance improvement*

Since DLP technology does not use polarized light, the mechanisms limiting contrast are completely different from those of liquid-crystal technologies, e.g. birefringence, skew rays, compensation films, etc. Instead, DLP contrast is limited by the geometrical aspects of the projection lens pupil, the pupil light distribution, scattering, and diffraction effects [HOR 99]. To improve contrast, up until 2001, efforts were concentrated on the device itself rather than the optical system in which it was used. Figure I.18 illustrates the evolution of mirror design to improve aperture ratio and therefore contrast [HOR 99].

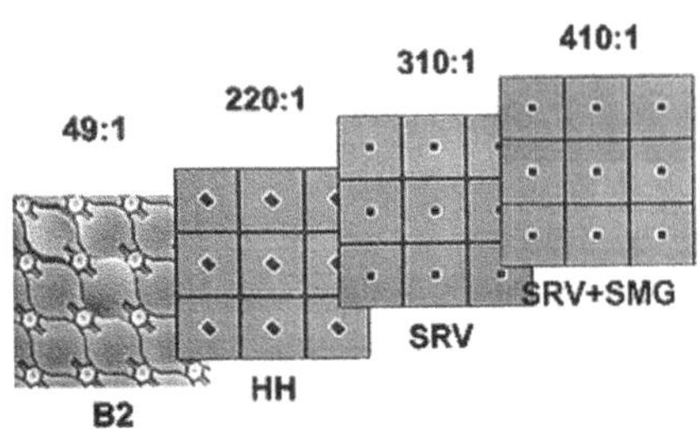

Figure I.18. *DMD pixel evolution, 1992–1997 to improve contrast. Reprinted with courtesy of the Society for Information Display*

Today, DLP displays offer excellent performance. This technology is now used in a great variety of projectors from very compact ones to professional, very high resolution and brightness systems for cinema. For wearable devices, the smallest size features 0.3 in DLP, with 720 p resolution. At the other end, DLP has been a key device to the rise of digital cinema, and are integrated in most cinema projector manufacture. Initial version of the device featured 2K resolution (2.2 million mirrors), but in 2012 4K versions (8.8 million mirrors) were proposed. Such projector use light sources with power of 1 kW and more. Typical device size varies between 0.98 and 1.38 in. in diagonal [BAR 14].

LCOS, transmissive LCD and DLP are today mature microdisplay technologies that are significantly present on the market, with a great presence in projection applications.

I.3.3.5. *Laser-beam scanners (LBS)*

Recently, new MEMS-based devices for display applications have been proposed. They consist of a moving mirror reflecting a laser beam on a projection surface. The mirror movement provides a scan on the projection surface, making an image. These devices, which can be used for compact projectors, are actually neither displays nor microdisplays since the image is not made on the device. Therefore, they are not within this book's focus. However, it is worth mentioning them since they can be used for the same applications than LCOS or DLP microdisplays.

I.4. New requirements for new applications: near-to-eye

Near-to-eye (NTE) consists of one (or several) microdisplay and an optical system used to create a virtual image for the eye. Such applications include head-mounted displays (HMD), viewfinders for cameras/camcorders, video goggles, etc. The recent rise of NTE applications pointed out the need for new kind of microdisplay.

All NTE systems need to combine the following features:

– Lightweight/compactness: NTE systems are wearable (HMD, video goggles) or present on portable devices (Cameras, cellphones, etc.), therefore they need to be as light and compact as possible.

– Low-power consumption: together with lightweight comes the problem of battery life and as a consequence the need for NTE systems to feature the lowest power consumption.

Microdisplays technologies such as LCD or DLP type are both non-emissive type i.e. they require separate light source such as arc bulb or LEDs or laser. As shown previously, requiring a light source means also requiring a dedicated optical system to collimate it onto the display. Both involve some additional volume and a more complex display system. Also, collimating an optical system might lead to loss in efficiency and lead to additional power consumption. For these reasons non-emissive displays are not ideally suited for NTE systems.

Therefore, with the rise of NTE the need for a self-emitting microdisplay made more and more sense in applications.

Self-emitting displays have existed for many years. In fact, from the beginning, since CRTs are self-emitting displays. Plasma displays are also of this type, which is also the case for field-emission displays. This last type, which appeared in the late 1980s and 1990s, consist of an array of individual electron sources and work on the same principle as CRTs. Here we do not consider vacuum-fluorescent devices (VFD), since their use is limited to alphanumeric display, not image. However, from the three relevant technologies (among which only plasma still exists today) it is not possible to shrink-down the size to fabricate microdisplays.

The most straightforward way to emit light is to make a light-emitting diode (LED). Starting from a CMOS active-matrix, it is logical to imagine such a device made from the silicon material. However, it is not possible with silicon to produce light sources, since its optical band gap is indirect. Therefore, Si LEDs cannot be integrated directly on the chip. These functions are reserved for other semiconductor materials, having direct gap, which are typically those from columns 3 and 5 of the periodic table. Devices made with such materials generally require stacks of epitaxial layers with appropriate thickness and doping profiles to build a light-emitting structure such as PiN junction, some of them including multiple quantum wells (MQWs). These materials are generally grown at quite high temperature (~700 °C or more), which means they cannot be built on an existing CMOS active matrix, which does not withstand temperatures higher than ~400°C. In such conditions, to make the full display, we need to couple the light-emitting array and the CMOS active matrix, which might be obtained with hybridization. This adds some inevitable costs to the fabrication process, together with potential difficulty to make as many connections as the pixel number.

For this reason, a direct deposition of light sources onto the CMOS circuits (monolithic integration) turns out to be a very attractive alternative.

I.5. Then came OLED microdisplays

Organic light-emitting diodes (OLED) became the first credible solution for efficient, potentially low-cost (monolithic integration), self-emitting display microdisplay.

The first OLED microdisplays were really developed in the late 1990s when researchers from several institutes and companies like the University of Arizona [MAT 99] and IBM [SAN 01] who published first CMOS-driven OLED display demonstrators. But the real pioneering company in OLED microdisplays is eMagin, which was created in 1996 in the US and rapidly showed demonstrators [JON 01, PRA 01].

The commercial stage of OLED microdisplay was achieved by eMagin in 2001 with the market release of the SVGA+ product [GOS 02, PRA 02] featuring full-color, low-voltage and high-brightness performance for near-to-eye applications.

MicroEmissive Displays (MED) was founded in the UK in 1999 with the aim of developing and commercializing OLED microdisplays targeting video glasses/personal video viewers, electronic viewfinders in digital cameras, camcorders and night vision systems. MED proposed polymer OLED (P-OLED) instead of the classical small-molecule OLED approach, giving a potential lower cost process.

Later, new actors such as MicroOled (France), Sony (Japan) and OlighTek (China) have emerged and sell OLED microdisplays.

Today, OLED microdisplays are widespread and have made their place in the display industry. They are manufactured by numerous companies, and are more and more present in applications such as video glasses (personal video viewers), electronic viewfinders in digital cameras camcorders and night vision systems.

Their expansion is not expected to stop. The combination of very low power consumption and compactness is making OLED microdisplays one of the best candidate for the growing demand of mobile, wearable devices.

I.6. Interactive glasses: the next big thing?

The announcement of the Google Glass in 2013 (Figure I.19) has raised considerable and undisrupted attention and hype since then.

This product appears to be the first step toward see-through, virtual reality glasses. Indeed, many other companies are now proposing see-through glasses, often with one system for each eye. This new kind of product has multiple potential applications which are yet to be defined. In all cases it could bring a revolution in our lives: after the keyboard and the touch panel, it could become the next-generation interface. Such interactive, see-through systems naturally bring new challenges for the displays together with tremendous new potential markets. Will OLED microdisplays take part in this revolution?

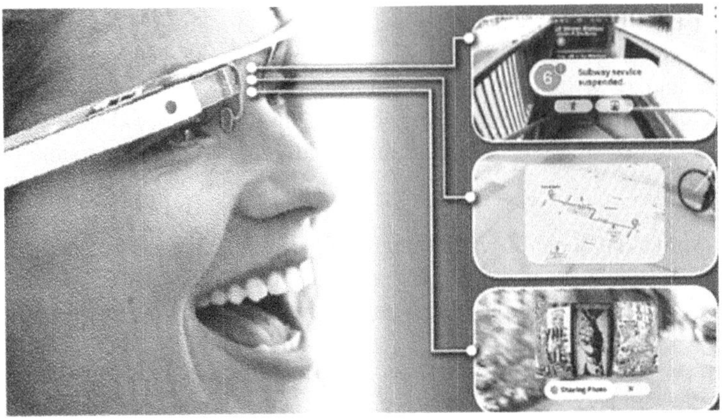

Figure I.19. *Google glass (source: Google)*

In summary, OLED microdisplays are now real actors on the display market, they are addressing more and more applications, and new applications are appearing.

I.7. Outline of the book

This book intends first to explain and describe how OLED microdisplays work and how they are made. But we will not go directly to this task. Indeed, Chapter 1 and 2 will first set the scene.

Chapter 1 aims to describe the basics of OLEDs: fundamentals, structures, materials, and basic device operation. Chapter 2 will try to give an overview of existing OLED-based products, which are now quite widespread around us.

Chapter 3 is dedicated to the characterization of OLED, i.e. how we can assess the main objective of such devices: the efficient conversion of electrical to radiated/optical power. Being able to predict their characteristics is a very useful feature for designing OLED devices, including OLED microdisplays: optical, electrical and electro-optical simulations allow it, as will be explained in Chapter 4.

The way the images are written on OLED displays, namely display addressing, will be detailed in Chapter 5, which will also include a description on how active-matrices are designed, from the pixel to the display level.

A major part of this book will be found in Chapter 6 where the fabrication of OLED microdisplays is detailed, from the active matrix to the final packaging and the intermediate steps of OLED deposition, encapsulation, and electronics.

The other objective of this book is to describe how OLED microdisplays are used today. This is the object of Chapter 7, where we review the two families of applications using the advantages of such displays: head-mounted displays, including informative glasses, and electronic viewfinder applications.

Also, and finally, we will explain what the expected evolutions of OLED microdisplays are with respect to the requirements of the new raising applications. This is the objective of Chapter 8. All of the figures in this book are available in full color at www.iste.co.uk/templier/oled.zip.

I.8. Bibliography

[BAR 14] Http//: www.barco.com. Accessed 2014.

[DE 02] DE SMET H., KUIPERS D., VAN CALSTER A., *et al.*, "Design, fabrication and evaluation of high-performance XGA VAN-LCOS microdisplay", *Displays*, vol. 23, pp. 86–98, 2002.

[DOH 98] DOHERTY D., HEWLETT G., "Phased reset timing for improved digital micromirror device (DMD) brightness", *SID'98, Digest*, pp. 125–128, 1998.

[GAN 00] GANDHI J., DAVIS D.J., JI Y., *et al.*, "Comparison of process tolerance of various LCOS modes", *Proceeding of the 20th International Display Research Conference*, pp. 233–236, 2000.

[GOS 02] GOSH A., "Full-color OLED on silicon microdisplay", *Proceeding of SPIE*, vol. 4464, pp. 1–10, 2002.

[HOR 89] HORNBECK L.J., "Deformable mirror spatial light modulators", *SPIE Critical Review Series*, vol. 1150, pp. 86–102, 1989.

[HOR 99] HORNBECK L.J., "A digital light processing update", *IS&T/SPIE Projection Displays V*, p. 158, January 1999.

[JON 01] JONES G.W., "Active-matrix OLED microdisplays", *SID Digest*, vol. 32, pp. 134–7, 2001.

[LIP 77] LIPTON L., STEPHENS C.P., LLOYD R., *et al.*, "A fully integrated MOS liquid crystal video-rate display panel with integrated drivers", *SID Digest*, vol. 8, p. 64, 1977.

[MAT 99] MATHINE D.L., WOO H.S., HE W., *et al.*, "Organic LEDs heterogeneously integrated with CMOS circuitry", *Digest of LEOS 12th annual meeting*, pp. 216–14, 1999.

[MEL 98] MELCHER R.L., OHHATA M., ENAMI K., "High-information-content projection display based on reflective LC on silicon light valves", *SID'98, Digest*, pp. 25–28, 1998.

[MOR 86] MOROZUMI S., SONEHARA T., KAMAKURA H., *et al.*, "LCD full-color video projector", *SID Digest XVII*, vol. 12, pp. 375–378, 1986.

[NTA 14] Http://ntact.com/wp-content/uploads/FPD-Revenues.jpg. Accessed 2014.

[PUT 10] PUTMAN P.H., "3LCD: projection technology for the 21st century", White Papers, Available at http://www.3lcd.com/explore/papers.aspx. 2010.

[PRA 01] PRACHE O., "Active-matrix molecular OLED Microdisplays", *Displays*, vol. 22, pp. 49–56, 2002.

[PRA 02] PRACHE O., "Full-color SVGA+OLED-on-silicon microdisplay", *Journal of SID*, vol. 10, pp. 133–138, 2002.

[SAN 01] SANFORD J.L., SANFORD J.L., SCHLIG E., *et al.*, "Direct-view active-matrix VGA OLED-on-crystalline-silicon display", *SID International Symposium Digest XXXIII*, pp. 376–379, 2001.

[SCO 03] SCOTT DEWALD D., SEGLER D.J., PENN S.M., "Advances in contrast enhancement for DLP projection displays", *Journal of the SID*, vol. 11, no. 1, pp. 177–81, 2003.

[TAN 07] TANAKA T., ITO A., YASUKAWA M., "New technology for HTPS-LCD panels for projection systems", *Journal of the SID*, vol. 15, no. 10, pp. 825–828, 2007.

1

OLED: Theory and Principles

The OLED acronym stands for organic light-emitting diode. It is a technology based on the integration of conjugated organic semiconductors, as thin solid films, in optoelectronic devices that emit light when electrical charge passes through by means of an applied potential. Today, this technology is mature for making displays and can be found in modern smartphones such as those manufactured by Samsung (Galaxy S family, see Figure 1.1 below).

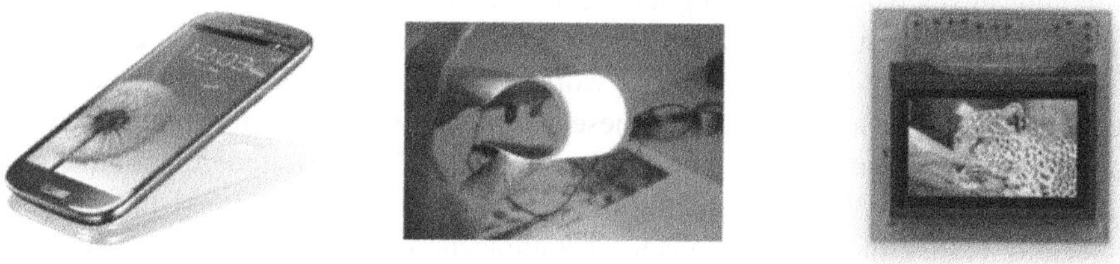

a) b) c)

Figure 1.1. *a) Samsung Galaxy S4 with an OLED display 1080 x 1920, 440 ppi (source: Samsung); b) flexible lighting fixture (source: LG); c) OLED-based Microdisplay 854 x 480 (source: MicroOLED). This, and all other figures in this book, is available in full color at www.iste.co.uk/templier/OLED.zip*

Another interesting application of OLED concerns the lighting market where they can replace their inorganic counterparts as the surface emitting medium instead of the point source emission light-emitting diode (LED). Finally, this book will address a niche application of OLED, which is microdisplay technology. Microdisplays are typical <1 cm^2 displays with a very high resolution. The market

---

Chapter written by Tony MAINDRON.

for microdisplays is currently undergoing rapid growth through new applications such as head-mounted displays (HMDs), a market that is now boosted by new three-dimensional (3D) applications, new concepts, such as informative eyewear, the replacement of optical viewfinders by electronic viewfinders in high-end digital cameras, as well as a broad range of high value professional applications such as digital night vision, medical applications such as vision aids, or industrial applications, e.g. in logistics and maintenance. According to a study of McLaughlin Consulting Group, the total market for microdisplays was around $500 M in 2012 with an average growth rate of more than 100% per year.

In this chapter, we will introduce a brief history of OLED and describe the basics of OLED, including processing methods and theory. We will emphasize OLED aspects that are at most related to microdisplay applications. A part of this chapter will be dedicated to thin-film encapsulation, which is an inherent characteristic of OLED-based microdisplays.

1.1. Organic light-emitting device: a brief history

In 1955, A. Bernanose presented a work related to the electroluminescence of organic compounds, such as gonacrin or carbazole [BER 95]. In 1963, Pope and his team published a paper addressing the electroluminescence of anthracene crystals. It was the first attempt to make an optoelectronic process onto an organic material-based device [POP 63]. However, this work on anthracene crystals remained at the fundamental research level. Indeed, these setups required much too high a driving voltage (>300 V) and were not reliable enough to be exploited for commercial electronic devices. At the beginning of the 80s, scientists [ROB 79] managed to noticeably reduce the driving voltage of anthracene-based devices. They used thin layer deposition techniques to spread this molecular crystal into thin amorphous layers. Nonetheless, device efficiency was low, too low to consider commercial exploitation of some system. It would take at least 10 years for the organic revolution to finally take place in optoelectronics.

In 1987, a Kodak team, initially dealing with solar photoreceptors, worked on an electroluminescent device with materials coming from xerography [TAN 87]. Indeed, they worked out the first organic electroluminescent diode with two sublimed juxtaposed thin layers. The stack, almost the same in use today, was as follows: indium tin oxide (ITO)/diamine/AlQ_3/MgAg. It was the birth of the OLED, now frequently used to refer to small molecule electroluminescent organic diodes. In these early devices, the two different layers have distinct electrical properties: the first layer, made up of a diamine (TPD standing for (N,N'-diphenyl-N,N'-di-(m-tolyl)-p-benzidine)) was a hole transport molecule and the second layer, AlQ_3 (standing for tris (8-hydroxyquinoline)aluminum) layer, was an electron transport molecule and the luminescent centre. Then, in 1990, a British team achieved an electroluminescent diode similar to the diode from Kodak using a conjugated polymer instead, poly (p-phenylene vinylene) [BUR 90]. This material simultaneously works as an electric charge bipolar carrier and a luminescent centre. Then, the PLED (polymeric light-emitting diode) was born. It is now commonly used and it more specifically designates electroluminescent devices made up of conjugated polymers instead of those made up of small molecules. The general OLED acronym remains the preferred term today to describe this technology from a general point of view.

1.2. Principles of OLED operation

In an OLED device, charges are electrically generated by applying a potential to the structure electrodes. Hence, considering structures made up of either sublimed small molecules or solution-processed polymer process leading to light extraction by luminescence may be described as follows: (1) negative (respectively, positive) charges injection in the electron (respectively, hole) transport layer by application of an electric potential to the respective electrodes; (2) electrical charges transport across the above mentioned layers, thanks to respective radical ions (oxydo reduction reactions); (3) coulomb capture of two opposite charges inside the luminescent material; (4) formation of an exciton on a luminescent material molecule, the excess of energy coming from the recombination allows the excitation of a molecule; (v) electromagnetic radiation due to the return of the molecule to its fundamental state. The emitted photon wavelength then depends on the material. For example, with AlQ_3 green light around 520 nm is emitted. Figure 1.2 depicts a generic OLED structure and a red-emitting OLED on glass substrate.

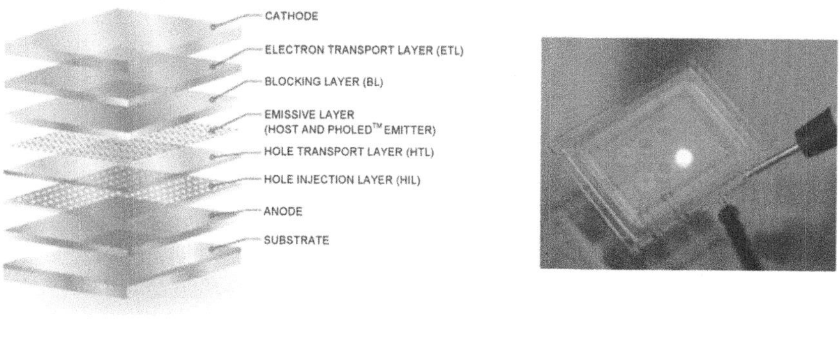

a) b)

Figure 1.2. *a) OLED typical stack (source: Universal Display Corporation); b) Red OLED emitting at 5 V (source: CEA-LETI)*

The manufacturing of organic flat screens and lighting sources depends on the ability to make the three primary colors, red, green and blue, which are the basis of these devices. These colors must be saturated and remain so, at least during the announced lifetime of the component. This implies a good temporal stability of the chromophores and, hence, a good electrochemical stability. White organic emission is usually performed by OLED exhibiting two or three primary chromophores in the same structure. Today the main challenge is the stability of OLED embedded organic materials. These devices must achieve high lifetime so as to compete with inorganic semiconductor devices, such as III-V compounds. The latter are indeed very promising for lighting purpose due to increasing efficiency (>100 Lm/W today) in the full visible spectrum. OLED long lifetime may only be reached by selecting high purity materials (to that sense monomers or small molecules are preferred today), careful device processing under controlled atmosphere and robust encapsulation against atmosphere. The final goal is to prevent chromophores (color source structural unit inside the material) from decay due to combined effect of light, oxygen or moisture.

1.3. Organic semiconductor material categories

1.3.1. *Small molecules*

Small molecule OLED technology has been developed by the Eastman Kodak company. Figure 1.3 shows detailed representations of some small conjugated molecules. From left to right AlQ_3, known for its electron conducting capacities and green visible photons emission, TPD: a hole carrier, DPVBi: blue emitter, DCM_2: red emitter and Irppy.

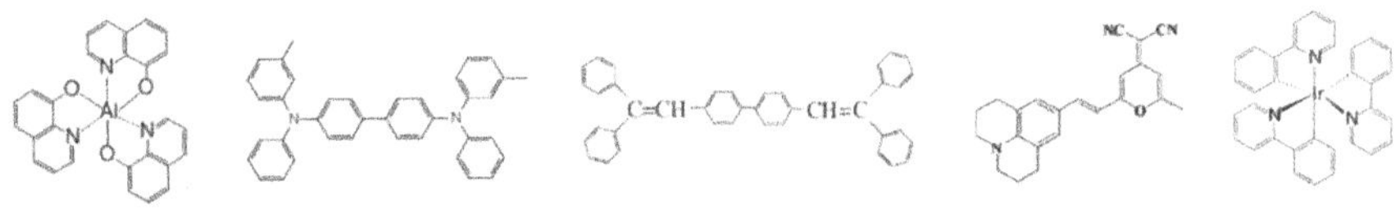

a) b) c) d) e)

Figure 1.3. *Small molecules examples. a) AlQ_3 electron transport molecule and fluorescent green emitter, b) TPD hole transport molecule, c) DPVBi blue emitter, d) DCM_2 fluorescent red emitter and e) IrPPy phosphorescent green emitter*

Small molecules are characterized by a low molecular mass. Optical and electrical properties can be optimized by judicious choice of the groups on the periphery (ligands). Thin layer deposition is obtained by simple vacuum evaporation or even wet deposition.

1.3.2. *Polymers*

By comparison, polymers have long molecular chains and hence a greater molecular mass. Several forms may be encountered: homopolymers, copolymers and dendrimers. Homopolymers are composed of a monomer chain, whereas copolymers result from the association of two or more different monomers repeated several times. Electrical and optical properties then depend on the type of monomers and the periphery groups, as shown in Figure 1.4 describing poly (phenylenevinylene) (PPV) and polyfluorene variants.

a) b) c) d)

Figure 1.4. *a) PPV electron and hole transport polymer and fluorescent green emitter, b) poly [2-methoxy-5-(2'-ethyl-hexyloxy)-1,4-phenylenevinylene (MEH-PPV)] electron transport molecule, c) polyfluorene (PFO) blue emitter and d) polythiophene hole transport polymer*

Electronic transport in polymers is similar to that of semiconductors since charges can delocate themselves on highest occupied molecular orbital (HOMO) and lowest unoccupied molecular orbital (LUMO) levels along the polymer chain (intramolecular conductivity). Nevertheless, like small molecules, charges can jump (hopping model) from one monomer on a chain to an adjacent monomer of another chain; hence polymers have a second type of conductivity, weaker than the first described and called intermolecular conductivity. Polymers are wet processed by spin coating or another variant of wet deposition process (ink-jet printing and screen printing). Today, this kind of process is also applied to small molecules.

1.3.3. *Deposition technique description*

OLED thin films, whose thickness does not exceed 100 nm, must be homogeneous and free from pinholes. The ability to be well deposited in thin layer is hence the sine qua non condition for these materials. For small conjugated molecules, it implies the possibility of vacuum evaporation without decomposition during heating (meaning that these fragile materials have a high vapor pressure). For polymers, they must be soluble in standard solvents to be deposited by spin coating or ink-jet method. Polymers cannot be deposited by vacuum evaporation because their molar mass is too high and they decompose in the crucible before achieving a vapor phase.

– Small molecules: vacuum evaporation

This process under vacuum leads to organic thin layers with very few impurities because the process itself is selective to the different monomers that are present in the crucible. Usually, train sublimation is employed to purify small molecules before their use in thermal evaporator. Impurities may contribute to an accelerated decay of devices or to a decrease of performances. Small molecules are far easier to purify and this is one of the reasons why they dominate on the market today (almost 100% of market share is attributed to small molecules-based displays). Polymers, on the contrary, are difficult to purify and their intrinsic nature, in particular their molar mass is polydispersed because of chain distribution, leads to reproducibility issues when mass production is concerned. However, small molecule thin films cannot be structured by classical photolithography techniques. Patterning is achieved with the use of shadow masking techniques in the vacuum tool (Figure 1.5).

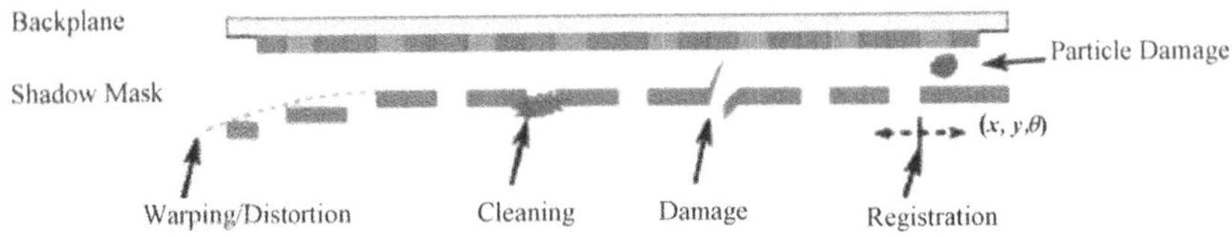

Figure 1.5. *Description of issues related to the use of shadow masks in vacuum tools for the deposition of small molecules (source: Dupont)*

It is generally a metallic plate perforated with the desired shaped holes so that the molecular vapor is replicated onto the substrate through the holes. This microstructuration method implies an alignment between the mask and the substrate inside the deposition chamber. It is limiting for large substrates processing or for high-resolution patterns because the masks tend to warp, and particle contamination occurs frequently. For OLED-based lighting sources and large display panels, increasing substrate dimensions will be a limiting factor. For microdisplays, the use of shadow masks is, to some extent, not a limitation because of the small size of substrate (silicon, 8) that contains the complementary metal-oxide-semiconductor (CMOS) driving circuits where the organic layers are to be deposited. Therefore, the fine patterning is realized by the CMOS pattern itself and the OLED deposited upon is a white-emitting OLED filtered out by means of color filters.

– Polymers wet deposition techniques

The main drawback of wet deposition process is the integration of impurities inside the film due to the not-under vacuum process conditions. It then requires perfect control of solvent purification (use of anhydrous solvent). This is not an easy task for large scale fabrication. That is why polymers are slightly behind small molecules from a commercialization point of view. The most frequently used deposition technique for polymers is spin coating. This technique is really easy to implement and allows deposition on large substrates at lower cost than small molecules. However, like for small molecules, there is no photo-microstructuration technique for these thin layers. However, the group of Meerholz in Cologne had presented in the past a family of light-emitting polymers that were photopatternable. They implemented these materials for patterning Red, Green, Blue (RGB) micropixels for microdisplays. The principle has been depicted in Figure 1.6.

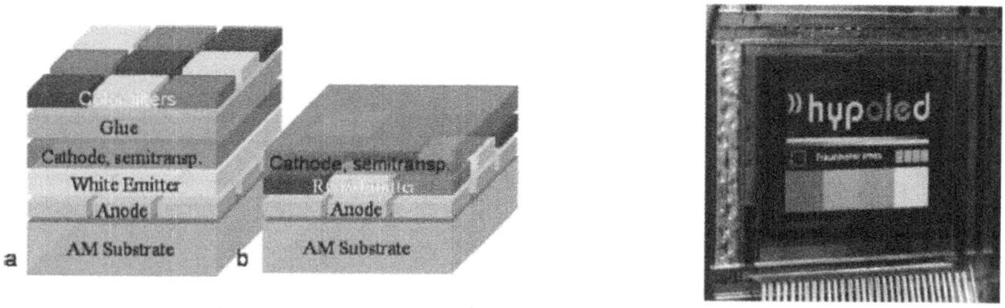

Figure 1.6. *Microdisplay principle using RGB photopatternable polymers (source: Hypoled project)*

To compensate for the lack of microstructuration techniques for organic semiconductors, ink-jet technologies or screen printing processes have been developed. The ink-jet process uses a printer and polymer in solution. Printer noses are adapted to fill exactly a tank which is usually surrounded by a hydrophobic barrier to prevent adjoining pixels contamination. However, for microdisplays, these deposition techniques are not suitable. Because of the tiny pixel size (< 10 μm), it is impossible to deposit RGB pixels side-by-side. If the microdisplay is made up of a white-emitting OLED in which light is filtered out by means of RGB color filters (Figure 1.6) solution-processed polymers or small molecules should be applicable. In a recent past, the UK company MicroEmissive Display (MED) had developed such a polymer-based microdisplay.

1.4. Organic semiconductors: theory

1.4.1. *Band theory in organic chemistry*

Semiconductor properties arise from the presence of a forbidden band gap in their electronic structure. This applies to organic semiconductors too. To understand the type of bond and the electric properties from a chemical point of view, let us study the case of a small amount of metallic atoms coming into interaction. When two atoms get close enough, their atomic orbitals interacts (Figure 1.7, case for ethane (C_2H_4)). It means that their wave functions interact too. Finally, one of the resulting wave functions has a greater energy (say antibonding) and the other function has a lower energy (say bonding). Consequently, the interaction between *n* atoms will generate *n* very close energy levels, some of them being bonding and some antibonding (the whole set is called a band).

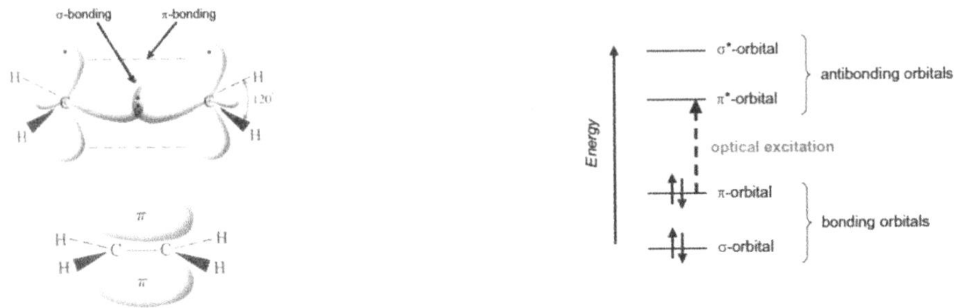

Figure 1.7. *Bond formation between two hydrogen atoms and consequences on electronic levels (source http://www.physik.uni-augsburg.de/organic/Dateien/OrganicSemiconductors.pdf)*

In a metal, the band will be half filled: the most stable energy levels being doubly filled whereas the other half higher energy levels being empty. This is only true for absolute zero observation. For other temperatures, many levels are partially filled due to Boltzmann distribution and energy level proximity; therefore, the limit is blurred. The highest energy level fully filled is called the Fermi's level.

1.4.2. *Differences from classical semiconductors*

A notable difference between organic and inorganic solids is that, in the case of organic solids, molecular entities are held together, thanks to Van der Waals interactions, whereas intramolecular bonds are covalent, respectively. Hence, the interactions between molecules are weak and the free mean path for electrons moving in the solid is roughly equal to the intermolecular space. As a result, the band structure of organic materials is localized, and not delocalized on the whole structure. For an isolated molecule, potential wells, induced by each nucleus, get together for the higher energies and form a large well. Deep atomic orbital are localized in the nucleus potential well, whereas higher atomic orbitals interact to form delocalized atomic orbitals, but only on the molecule. Two levels may be distinguished: the HOMO level and the LUMO level. They can both be compared to valence and conduction bands, respectively, for inorganic semiconductors, yet localized at the molecular level. Therefore, in organic solids, charge transport is incoherent. Every transport site is localized because energetic and positional disorders are largely greater than the coordinated structure. Hence, conductivities in organic solids stand far lower than in inorganic solids. From a chemical point of view, charge transportation occurs through an oxydo-reduction process between molecules. Molecular films such as the ones used in OLED technology are amorphous in nature.

As described in Figure 1.8, amorphous organic solids have two densities of localized levels merged into Gaussian forms. That explains the frequent use of Brownian motion laws or Monte Carlo algorithms to model charge transport. However, weak intermolecular bonds allow mechanical flexibility without penalizing defect formation. Consequently, fabrication of conformable displays or lighting devices becomes possible assuming a flexible substrate too (Figure 1.9).

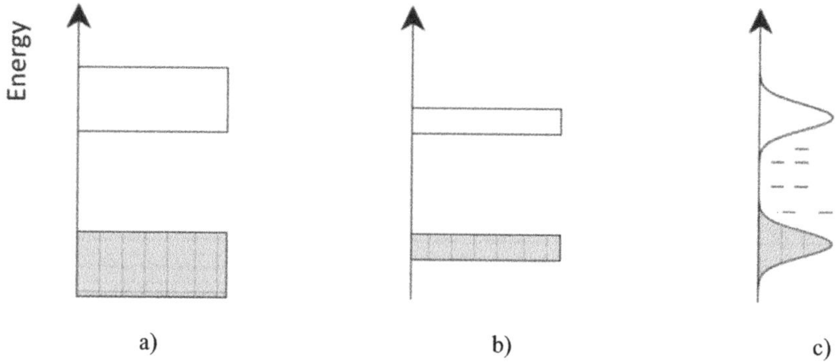

Figure 1.8. *a) Comparison between large electronic band structures in semiconductors, b) electronic levels in organized organic semiconductors and c) unorganized organic semiconductors*

Figure 1.9. *LG prototype of flexible display based on the integration of an Active Matrix OLED (AMOLED) – the flexibility of the organic material is made possible thanks to the weak Van der Waals interaction between molecules (source: LG)*

1.4.3. *Electronic transport model in amorphous organic solids*

In the case of amorphous molecular solids (which are the core of OLED devices) electronic transport is of the localized kind on the individual molecule. Usually, localized jump from molecule to molecule (hopping model) formalism is applied to describe charge transport. This model was initially developed to describe transport in solid such as amorphous silicon. This model describes a tunneling effect between molecules sustained by network vibration. It enabled description of charge carriers' movements inside unordered covalent crystals with numerous structural or chemical defects. It was first described by Conwell [CON 56] and then put into mathematical form by Miller and Abrahams [MIL 60]. The transition rate (number of tries per second) from a site, or a trap, *i* with an energy ε_i to another site *j* with an energy ε_j may be expressed as follows:

$$\nu_{ij} = \nu_0 \exp(-2\gamma_{ij}\Delta R_{ij}) \exp\left[-\frac{(\epsilon_j - \epsilon_i)}{kT}\right]$$

For $\varepsilon_j > \varepsilon_i$, and:

$$\nu_0 = \exp(-2\gamma_{ij}\Delta R_{ij})$$

For $\varepsilon_j < \varepsilon_i$

ν_0 is the phonon vibration frequency corrected by a probability of tunneling effect for jumps toward lower energy states ($\varepsilon_j < \varepsilon_i$). This probability depends on the intersite distance (ΔR_{ij}), a γ_{ij} term, representing the overlapping of the wave functions of states *i* and *j*. A Boltzmann distribution, which describes a phonon absorption probability, is added in the case of jumps toward higher energy ($\varepsilon_j > \varepsilon_i$). The formalism used to describe charge carriers mobility in organic materials depends on the nature and the structure of the material itself. The polaron formalism [GIL 72] is usually applied to conjugated polymers and molecular crystals films, which have a narrow band structure. The disorder theory or Gaussian Disorder Model (GDM) [BÄS 81] best describes the mobility behavior in amorphous organic glasses. In this formalism, a general model proposes a dependency of the charge mobility as a function of $E^{0.5}$ and $(T_0/T)^2$:

$$\mu(E,T) = \mu_0 \exp\left[-\left(\frac{2\sigma}{3kT}\right)^2\right] \exp\left\{C\left[\left(\frac{\sigma}{kT}\right)^2 - \Sigma^2\right]\sqrt{E}\right\}$$

Here the disorders in both position (including orientational effects) and energy are considered to be Gaussian distributions with widths of Σ and σ, respectively. *C* is an empirical constant, while just the temperature dependence (against a measure of the energetic disorder T_0) of the zero-field mobility is given by:

$$\mu_{E=0} = \mu_0 \exp\left[-\left(\frac{2\sigma}{3kT}\right)^2\right]$$

The disorder model turns out to be the most prominent model for fitting both the field and temperature dependence compared to well-known Poole–Frenkel model, which describes only the field dependency of the mobility. Today, derivatives of the GDM are known as Correlated Disorder Models (CDMs) or Extended Gaussian Disorder Models (EGDMs).

A general illustration of such a mobility behavior in amorphous solids could be found in Figure 1.10 [BOR 91]:

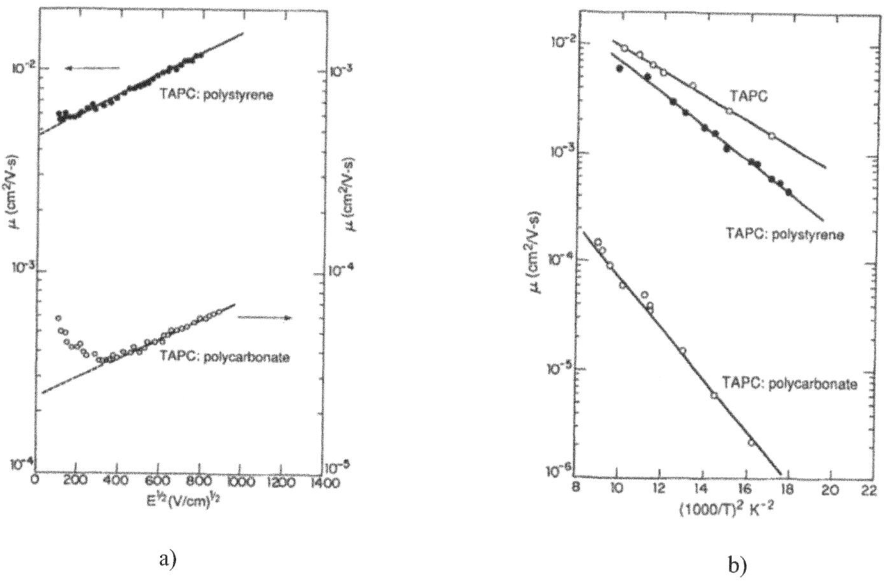

Figure 1.10. *a) Logarithm of the mobility versus $E^{0.5}$ for a hole-transporting molecule 4,4'-Cyclohexylidenebis[N,N-bis(4-methylphenyl)benzenamine] (TAPC) blended into a host polymer (measured at 295 K) and b) logarithm of the zero-field mobility versus T^{-2} for the same system*

These characteristics of amorphous solids well described by the above mentioned models will be part of the models developed for simulating current–voltage curves of OLED, as described below.

1.5. OLEDs electrical characteristics

Charge-carrier transport in amorphous semiconductors is not yet fully understood. A comprehensive theory for mobility and current–voltage curves is still needed to optimize the device performance. Models to explain charge transport in organic semiconductor fall into two regimes of operation: (1) the injection limited transport which supposes that the injection barrier between an electrode and an organic adjacent thin film is the main hurdle for charges to move from one side to the other; (2) the bulk limited transport which assumes that the supply of carriers into the layers is not the limiting factor so that the amorphous organic layer itself is the limiting factor for charge transport (see Figure 1.11).

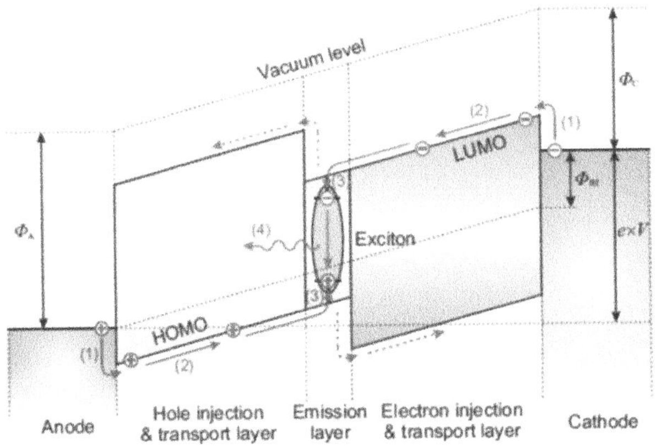

Figure 1.11. *Illustration of an OLED structure showing the two important mechanisms found in an OLED which are injection of charge and transport of charges (with the presence of energetic barriers at metal/organic interface) (source – [BRÜ 13])*

Former charge injection mechanisms have been directly taken from the conventional semiconductor theory as described by the models of tunneling (Fowler–Nordheim model) and thermionic emission (Richardson–Dushman equation) in the case of a contact between a metal and a semiconductor. The transport mechanisms in the amorphous solid material that have been proposed were: ohmic conduction and conduction limited by the formation of space charges with traps giving rise to trap-charge limited current (TCLC) or without energetic traps giving rise to space-charge limited current (SCLC).

Similar to the conventional semiconductor device (PN junction, for example), OLED consists of metal/organic interfaces. This type of contact arises as soon as the Fermi energy levels of the two materials are offset. Thus, when placed in intimate contact, the thermodynamic equilibrium involves a joint alignment of these levels at the cost of charge transfer at the interface. Researchers in the past have shown that neither the model fowler nordheim (FN) nor the typical equation of PN junctions can describe well the JV curves of OLED based on small molecules like TPD/Alq3 and this, regardless of the applied electric field [BUR 94]. Their comments were more in favor of a description dominated by the effects of space charge limited current. Space charge current models have been described in the early 50s for amorphous semiconductors. Then Mark and Helfrich introduced them for describing transport in organic crystals [MAR 62]. In that case, the formalism does not take into account the dependence of the mobility on the field and temperature because organic crystals are supposed to be free of intrinsic charges so that the net charge density that contributes to the whole current is an uncompensated charge density injected by the contact electrode. The law of steady-state current is often described by the law of Child or Mott–Gurney [MAR 62]:

$$J_{SCLC} = \frac{9}{8}\epsilon\mu\frac{V_{appl}^2}{d^3}$$

$\varepsilon_0\varepsilon_r = \varepsilon$ is the permittivity of the material, d is the thickness of the organic thin film, V_{appl} is the applied potential and μ is the carrier mobility.

For an amorphous semiconductor, energetic traps are present in the band gap of the material. As a result, current–voltage plots for such materials deviate from the V^2 law to higher exponent. Therefore, the new current regime is described in terms of TCLC. The net current is due to the motion of charge carriers trapped and thermally released by localized trap states in the band gap (band gap notion for amorphous semiconductors is a blur notion due to their amorphous form they possess a localized density of state (DOS) rather than energetic bands – however, the term is empirically used in the OLED community), and that the frequency of these trapping events is significantly high to limit the conduction in the film. Therefore, TCLC is a modified SCLC that includes a trap distribution. For undoped OLED (see section 1.6), at high field the Fermi level will move closer to the conduction DOS so that the energy levels of all the intrinsic traps in the amorphous solid will be filled. The further injected charges will then follow a trap-free conduction at high voltages. Note that if some intrinsic charges are thermally generated at low voltages, the ohmic conduction can be effective and the following expression applies:

$$J_\Omega = n_0 q\mu\frac{V_{appl}}{d}$$

Where n_0 stands for the free carrier density in the organic film. An example of a PPV-based OLED (referred as PLED in the past) has been depicted in Figure 1.12:

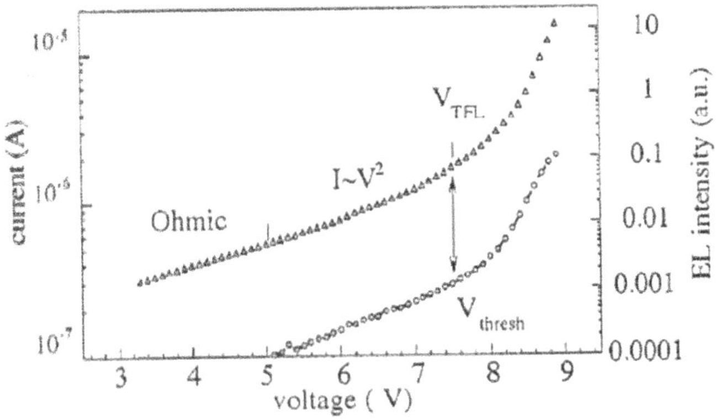

Figure 1.12. *Typical JVL characteristics of an OLED (source: [SAV 97]) showing the different regimes of space-charged limited current*

For an exponential distribution of trap states, described as followed:

$$n_t(E) = \left(\frac{N_t}{kT_t}\right) exp\left(\frac{E - E_C}{kT_t}\right)$$

where T_t, N_t and E_c are the temperature traps, the energy level of the conduction band and the total density of traps, respectively, the shape of the TCLC becomes [MAR 62]

$$J_{TFL} = N_C \mu q^{1-m} \left(\frac{m\varepsilon}{N_t(m+1)}\right)^m \left(\frac{2m+1}{m+1}\right)^{m+1} \left(\frac{V^{m+1}}{d^{2m+1}}\right)$$

N_C here represents the density of states of the conduction band, the exponent m is defined by $m = T_t/T$ and must always be greater than 1. The larger it is, the larger the exponential distribution $n_t(E)$ decreases slowly and the more there are traps.

In the early 2000s PIN OLED were introduced to show high efficiencies at very low voltages. These architectures are made up of electrically p-doped and n-doped layers contacting the interface electrodes similar to PN junction. For these devices (see section 1.6), current–voltage formalism could be a bit different. Recently, a model has been proposed to fit the current–voltage characteristics based on an analytical mixture of a SCLC term with an exponential term that resembles the current characteristics of inorganic PN junction:

$$I3 = c_3(V - V_{BI})^{n3} + c_4\left[exp\left(\frac{V - V_{BI}}{c_5}\right) - 1\right]^{n4}$$

These last devices are used in OLED-based microdisplays.

1.6. OLED: different structure types

1.6.1. *Direct and inverted diodes*

Starting from the substrate, OLED layer stacks can start with either an anode or a cathode; they are then called direct diode or inverted diode, respectively. The choice of one of these two stacks has consequences on OLED electrical addressing and on carrier injection in the structure. There are few examples of inverted diodes compared to direct diode. This is because cathode-organic semiconductor's good electrical interfaces are very difficult to achieve when a metallic cathode is of concern. This is because metallic cathodes tend to passivate their surface when exposed to O_2, making an insulating layer that impedes charge injection into the adjacent organic electron transporting layer (ETL). Electrical contact cathode onto organic semiconductors is, on the contrary, simpler to realize for direct structure because the evaporation process of the metal is energetic enough to allow shallow penetration of metal atoms in the soft organic matter. As a result, the organic/metal interface in these structures tends to be diffuse rather than sharp.

1.6.2. *Through substrate emitting diode and top surface emitting diode*

In an OLED, photons can be emitted through a transparent substrate (bottom-emitting OLED). But they can also be emitted through the top of the stack, usually through the cathode in a direct structure (top-emitting OLED). This choice implies a discussion on electrodes' type. Usually, for bottom-emitting diode, anode is composed of a transparent conductive oxide (TCO) such as ITO or indium zinc oxide (IZO).

Top-emitting OLED releases constraints on the kind of substrate. In a display, an addressing circuit is placed under the OLED stack to realize an active matrix display. In that case, a reflective base electrode is necessary so that the light can be reflected back to the top of the structure and extracted. In the case of microdisplays CMOS driving circuits are used as addressing backplanes while Al/TiN materials are used as the anode for the OLED made upon (Figure 1.13).

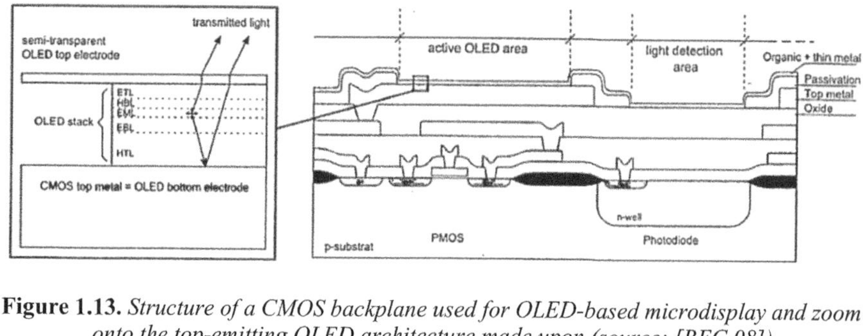

Figure 1.13. *Structure of a CMOS backplane used for OLED-based microdisplay and zoom onto the top-emitting OLED architecture made upon (source: [REC 08])*

1.6.3. *Heterojunction diode and band engineering*

– Electrode choice

To generate photons, negative and positive carriers must be first injected into the semiconductor material adjacent to the respective electrodes. Hence, an anode and a cathode have to be chosen suitably with a work function which limits the potential barrier to, respectively, the HOMO and LUMO level of the adjacent semiconductor. At least one electrode must be transparent or half transparent and both electrodes must be as much stable as possible to avoid premature degradation of the device.

– Quenching phenomenon

A single layer OLED shows limited efficiency because of quenching phenomenon at the electrode (mostly at the cathode because holes are 100 times more mobile compared to electrons), of conductivity difference between HOMO and LUMO levels of a given organic layer and also because the injection differences between holes and electrons at their respective electrodes imply the domination of a single type of carrier. This unbalanced current in the structure necessarily implies a poor efficiency for the single layer OLED. Today OLED structures have evolved toward multilayer architectures which use different organic semiconductors according to their major specificity. Associating a high electron conductivity semiconductor with the cathode and a high hole conductivity semiconductor with the anode allows the recombination zone to be fixed far from the electrodes, close to the heterojunction interface. Holes injection layer (HIL) and electrons injection layer (EIL) can also be added to the above list. They improve carrier injection from the electrodes to the transport materials. As a result, quenching phenomena at electrodes are impeded.

1.6.4. *Electrical doping*

The basic principle of electrical doping in organic semiconductors is the same as in inorganic semiconductors. Impurities are added which either transfer electron to the electron conducting band or states (LUMO) or remove an electron from the hole conducting band or states (HOMO) to generate a free hole. Some constraints appear for the host and the guest molecule. In the case of *p*-doping, the LUMO of the dopant, a strong electron acceptor, must be energetically below the HOMO of the host. This is the case for the well-known 2,3,5,6-Tetrafluoro-7,7,8,8-tetracyanoquinodimethane (F4-TCNQ) molecule for which electron affinity stands around 5.24 eV. When doped into a spiro-TTB hole transporting matrix (molecule used in OLED), for which HOMO lies around 5.2 eV, it can pull out electrons from the host molecule so as to release free holes. The resulting spiro-TTB film is then degenerated and its conductivity can be increased to several orders of magnitude. An example of such a conductivity increase has been depicted below for a system ZnPc:F4TCNQ (Figure 1.14).

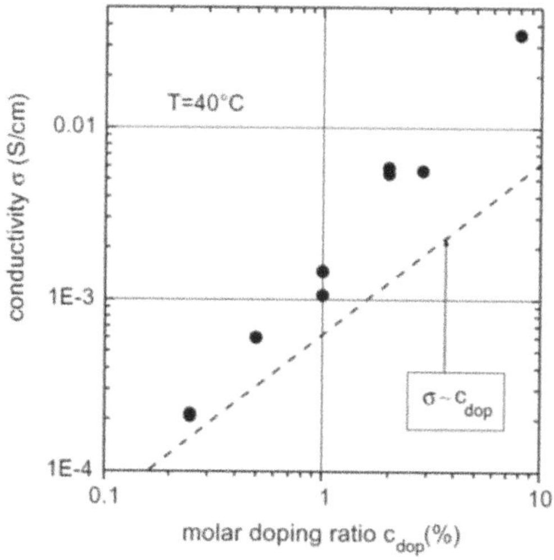

Figure 1.14. *Conductivity of p-doped ZnPc as a function of the doping concentration with the molecular dopant F4TCNQ (source: M. Pfeiffer et al. Org. Elec. 4 (2003) 89-103)*

For *n*-type doping, the HOMO of the dopant, a strong electron donor, must be energetically above the LUMO level of the host matrix. For that reason, molecular *n*-dopants are unstable to air in particular because the electron affinity of OLED *n*-type molecules lies around 3 eV. As an alternative to molecular dopant, we may use Li, LiF, Liq, Cs, Li_2CO_3 or $CsCO_3$ to dope the host matrix. The drawback of using such materials, however, is related to their diffusion inside the structure under applied electrical field or thermal activation [MAI 06]. As a result, the stability of these systems is quite relative.

The basic principle of electrical doping OLED is to achieve low voltage efficient OLED which is not limited by the injection at the contact. Usually PIN OLED is made up of two electrically doped layers adjacent to the respective electrodes, transporting layers for holes and electrons (undoped) which act as electron blocking and hole blocking layers, respectively. In the middle of the stack stands the emitter system. The use of highly doped transporting layers has two key advantages. The first advantage is the low electrical field in the layers because of the high conductivity. The OLED can operate in flat band conditions so that the operating energy is close to the photon energy (for a green device $eV = h\nu \rightarrow V \sim 2.5$ V). Second, the heavy doping allows us to make ohmic contacts so that the thin barrier formed by the space charge allows ohmic injection despite the high energetic barrier between the Fermi level of the electrodes and the respective transport DOS in the organic layers. PIN OLED shows very steep IV curves compared to undoped OLED which is very useful for display application and *a fortiori* microdisplays. An example of an IV characteristics of a PIN OLED versus an undoped OLED is presented below (Figure 1.15) [HUA 02].

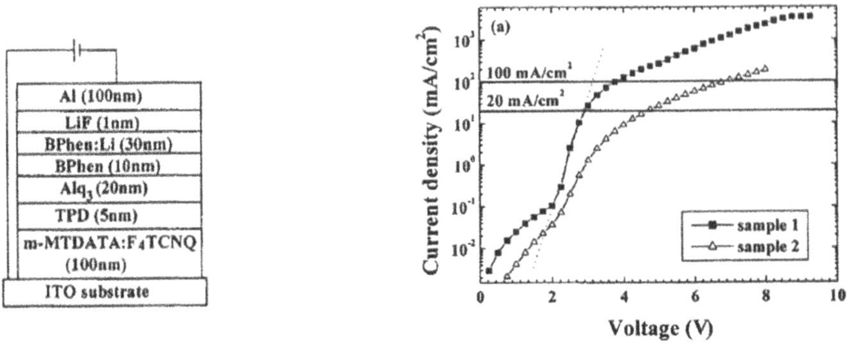

a) b)

Figure 1.15. *a) PIN OLED architecture; b) IV characteristics of the OLED depicted in a) with n-doping (sample 1) and without (sample 2)*

1.6.5. *Light extraction*

Organic materials usually have an optical refractive index n around 1.8. Thus, emitted photons from the emissive layers are mainly trapped in the guided modes because of the total internal reflection angle θ_c. In the case of organic materials from which light is extracted to the external world with $n = 1$ (air), this angle is equal to 36°. In the case of an isotropic source, the ratio of extracted light to the total flux, denoted η, could be expressed as follows:

$$\eta = \frac{\Omega_C}{\Omega} = \frac{2\pi.(1 - cos\theta_C)}{2\pi}$$

A numerical calculation gives $\eta = 19\%$ which corresponds to the amount of light extracted from an OLED. We can see that this number is very low. Different methods exist that optimize light extraction out of the devices. Usually people improve light emission by increasing the surface roughness of the substrate [WIN 99] from which light beams are escape or by adding silica microspheres [YAM 00], microlenses [MAD 00] or shaped mesa [GU 97]. Recently, the company Panasonic Electronics Works announced a world record of efficiency for a white bottom-emitting phosphorescent tandem OLED of 110 lm/W on a panel area of 25 cm² using an optimized light outcoupling architecture together with an all phosphorescent white OLED. The structure has been depicted in Figure 1.16.

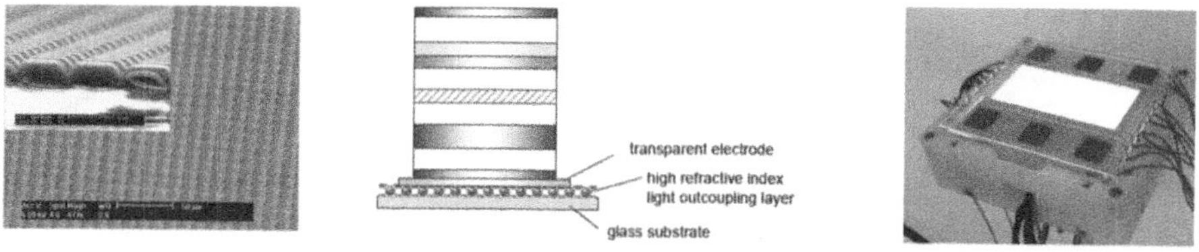

a) b) c)

Figure 1.16. *a) SEM image of microlenses (source: S. Möller and S. Forrest J. Appl. Phys. 91 (2001) 3324); b and c) white bottom-emitting phosphorescent tandem OLED developed by Panasonic using a high refractive index light outcoupling layer onto the glass substrate (achieving 110 lm/w for a 25 cm² emitting surface) – (source: [OSA 14])*

Understanding of interference phenomenon occurring in red, green and blue emitting OLED structures is the key to obtaining high efficiency. This is particularly true for top-emitting OLEDs. Two major types of interference mechanisms are presented in Figure 1.17. Their optimization will raise device outcoupling, that is to say photons ability to escape from the OLED. Direct interferences between the most reflecting electrode (usually the anode) and the emitter and the second electrode clearly illustrate the effects of the cavities (Fabry–Perrot type) generated in these structures.

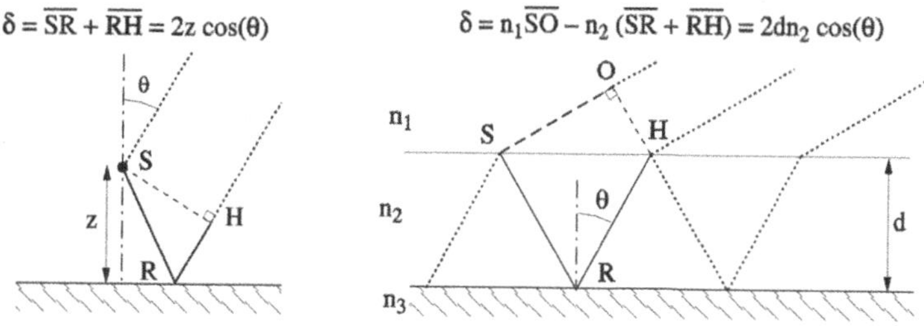

a) b)

Figure 1.17. *Schematic view of two types of encountered interferences in top-emitting OLED. a) Direct interferences and b) cavity effect*

The advantage of PIN OLED for top-emitting architectures seems quite obvious. Because the conductivity of the *p*-doped and *n*-doped layers is high and constant on a large thickness range, this means that the cavity of such devices can be tuned to the different cavity orders without ohmic loss in the structure. Following Figure 1.17, it can be said also that *z* and *d* can be tuned independently by adjusting hole transporting layer (HTL) and ETL thicknesses as described below on the simulation (Figure 1.18).

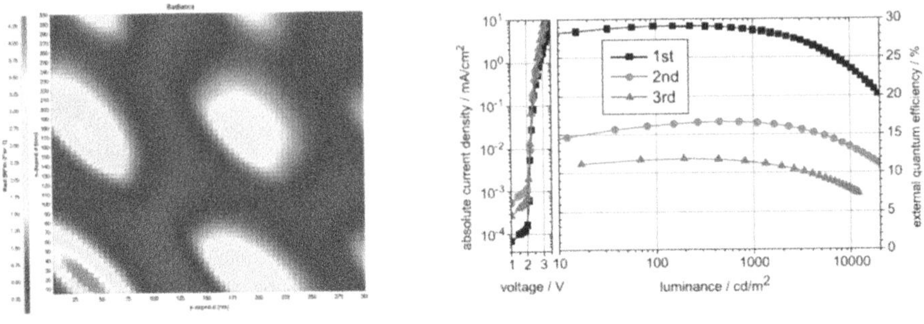

a) b)

Figure 1.18. *a) Optical simulation (using SETFOS software) of a PIN OLED showing the different cavity optima (radiance) versus p-doped and n-doped layer thicknesses; b) IV characteristics of a green OLED for first and second order cavity (source – [HOF 10])*

– Fluorescence versus phosphorescence

In an OLED, the luminescence is obtained by means of fluorescence or phosphorescence. In OLEDs using only fluorescent emitters, the radiative decay of triplet excitons (75%) is spin forbidden, so only the singlet excitons (25%) can emit light. Conversely, noble-metal-based organometallic phosphors exhibit emissive triplet states because of singlet–triplet mixing via effective spin–orbit coupling, so phosphorescent organic light-emitting diode (PHOLEDs) containing them can exhibit near-unity internal electroluminescence quantum efficiency.

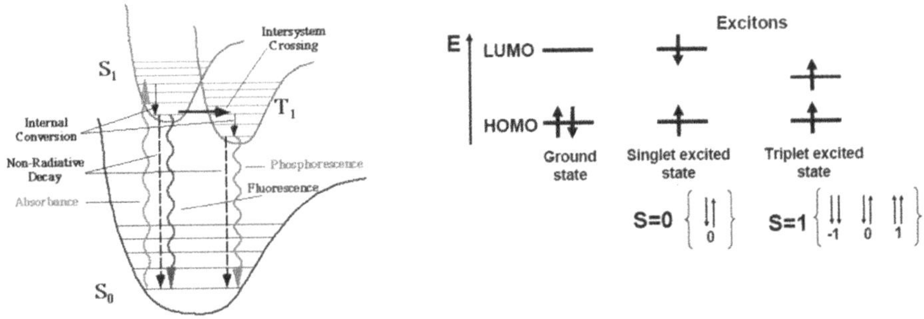

a) b)

Figure 1.19. *a) Band diagram (left) and efficiency (right) of two luminescent organic material a fluorescent material and b) a phosphorescent fluorescent material (source: Google)*

1.6.6. *OLED efficiency*

To understand and, thus, optimize the performance of an OLED, it is important to take into account the different mechanisms leading to loss of efficiency. There are four major mechanisms that are somehow related to each step of the operation of an OLED. Each can be represented by an efficiency, the product of four individual efficiencies giving the total efficiency is the external quantum efficiency. This overall yield is generally denoted as η_{ext}. It can be expressed mathematically as:

$$\eta_{ext} = R_e \eta_{int}$$

η_{int} representing the internal quantum efficiency, e.g. the efficiency that does not take into account the optical extraction from the device, or the emitted light. It could be written as the product of three quantum efficiencies:

$$\eta_{int} = \gamma \eta_S \Phi_f$$

where g, n and ϕ represent, respectively, the ratio of injected holes to injected electrons (charge balance) in the device, the singlet to triplet states ratio and luminescent efficiency (competition between radiative and non-radiative decay). Today the best internal efficiencies are close to 100%. By adding suitable optics for light outcoupling, it is possible to achieve high external quantum efficiencies. Recently, people from IBM released a paper with a green-emitting OLED that shows a 60% external quantum efficiency (Figure 1.20).

Green OLED IBM	Without extracting medium	With extracting medium
Luminance (cd/m²)	3000	10000
EQE (%)	20%	60%
Current efficiency (cd/A)	80 cd/A	240 cd/A

a)

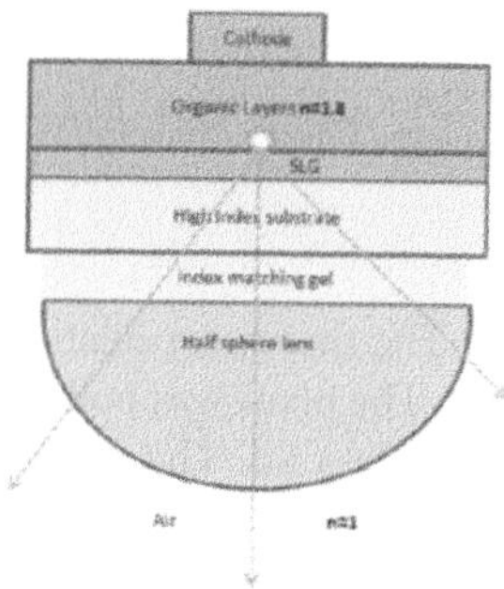

b)

Figure 1.20. *a) recent result obtained by an IBM team onto green bottom emission OLED; b) structure of the green OLED with the extracting medium (source [LI 13])*

1.7. OLED stability and lifetime: encapsulation issue

The issue with organic materials is their high sensitivity to gaseous elements in the atmosphere, particularly O_2 and H_2O. That is why their protection in an electroluminescent device is important to prevent accelerated decay. Generally a protection, or an encapsulation, of circuits is performed so as to protect active organic layers and sensitive cathode metal. Figure 1.21 describes the encountered problems with organic semiconductors and depicts the encapsulation concept for OLED (this concept is also valid for all electronic devices using hydrocarbonated materials, such as photovoltaic cells or transistors). As illustrated in the figure, it is mandatory to protect the organic molecules at each fabrication step of the assembly, that is to say from the raw material obtained by chemical synthesis (safeguarding of the powders under inert atmosphere, for example) to the commercial device (lamp or display) through the OLED made in the laboratory.

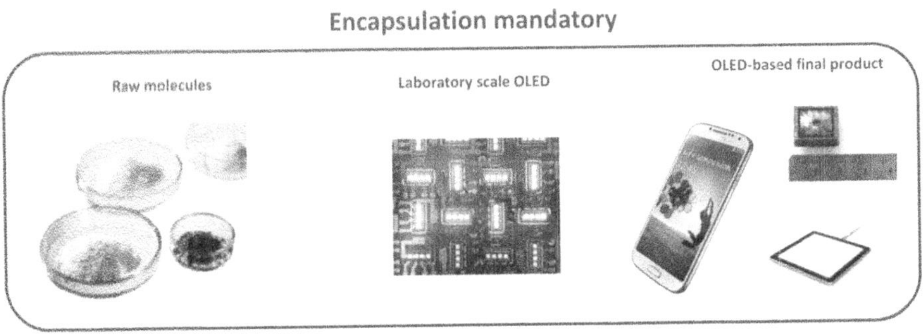

Figure 1.21. *Encapsulation issues for OLED devices from the raw materials to the products (source – Novaled, Philips, Samsung and Emagin)*

The first encapsulation concept of an organic diode was published in 1994 [BUR 94]. The pixel was built on a glass/ITO substrate. On the other side, it was protected by a glass lid. The assembly was held together, thanks to an epoxy resist. In 1996, this concept was patented by a pioneer who added a desiccant (P_2O_5) inside the space housing the device [Patent EP0776147, Figure 1.22]. Today it is still the preferred process that can be encountered in commercial products. Commercial products like the Samsung galaxy family are made this way with the lid being fritted to the glass substrate by mean of a laser frit technology. This ensures of course a highly hermetic cavity compared to a usual perimeter glued epoxy or acrylate polymer resin.

The concept of protection against oxidizing gas of atmosphere (H_2O, O_2) is mandatory for thin amorphous molecular films as well as for OLED. Usually, thin inorganic barrier layers are made up of oxides, nitrides or oxynitrides usually deposited by low temperature plasma-enhanced chemical vapor deposition (PECVD) and sputtering techniques [KIM 07]. Usually very low gas permeability level, water vapor transmission rate (WVTR) around 10^{-6} g/m²/day and oxygen transmission rate (OTR) around 10^{-3} sccm/m²/day could be reached by building hybrid encapsulation stacks of the kind [organic/inorganic]*n*, $n = 4$ on top of the device [BUR 01]. For that purpose, thick (~ 0.5–5 μm) vacuum deposited smoothing polymer layers are deposited between the thin mineral barrier layers (≤100 nm) which inevitably possess some intrinsic defects such as pinholes or cracks. Their role is to decouple these defects from one barrier layer to another so that the paths for the oxidizing species to reach the fragile device are lengthened, and the whole degradation is delayed [AFF 96].

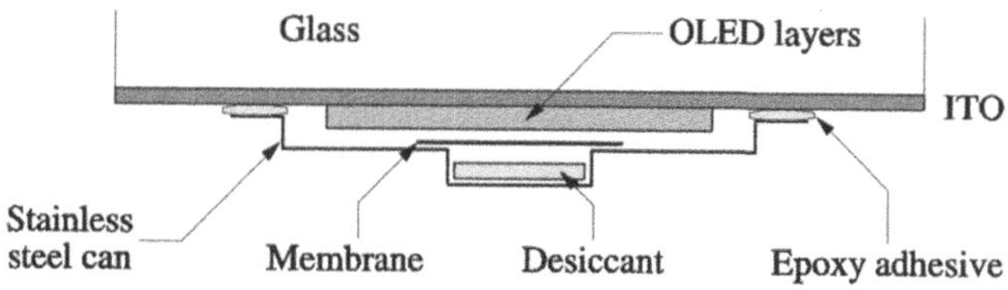

Figure 1.22. *OLED device first encapsulation concept (source – Pioneer 1996)*

The research and visualization of defects on the surface of transparent inorganic thin film coatings such as SiO_x, used for barrier applications against oxygen and water ingress, has been a hot subject of research for a long time. The food packaging industry has driven this research for many years (1–6) while the renewal of interest for this technology is related to the rapid development of organic electronics [FOR 00]. The levels of barrier quality against moisture ingress are, however, totally different between the two disciplines. For OLED in particular, moisture is at first order the main cause of degradation of devices and a typical WVTR as low as 10^{-6} g/m²/day has to be reached for OLED applications to ensure their commercial viability [BUR 01] (Figure 1.23).

The usual phase vapor deposition (PVD) sputtering or low temperature PECVD techniques do not allow the realization of single inorganic films that can fulfill this specification [CHA 96, SOB 98]. Thus, the search for ultrahigh barriers has led to the development of multilayer hybrid encapsulations and the emergence of deposition techniques like the atomic layer deposition (ALD) usually related to the deposition of single barrier film of mostly Al_2O_3 material [GEO 10, GRO 04], suitable for OLED-based microdisplay encapsulation.

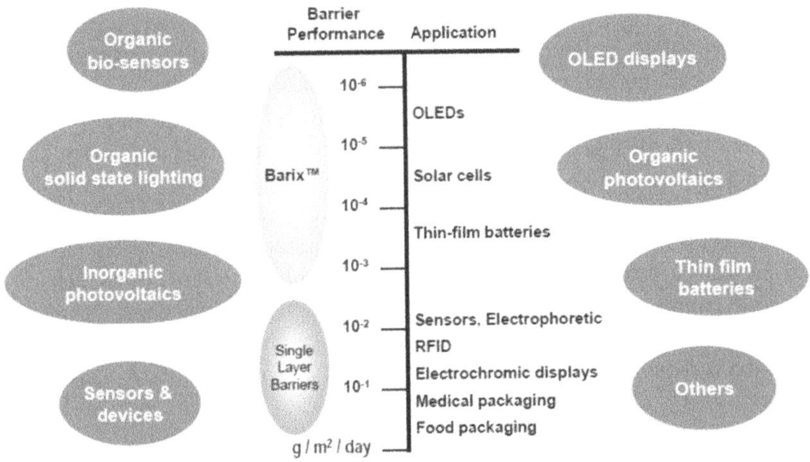

Figure 1.23. *Encapsulation levels (to water) versus technology needs (source: Vitex, 2008)*

This derivative of the chemical vapor deposition (CVD) technology is highly suitable for OLED because it allows the deposition of (almost) pinhole-free and highly conformal and reproducible, oxide films at temperatures that are compatible (100°C typically) with the limiting low transition temperature (T_g) of organic semiconductors that compose these devices. A couple of publications have demonstrated the capability of Al_2O_3 ALD films to reach this severe level of WVTR requirement [LAN 06, GRO 06, CAR 06] as well as their applicability to encapsulate OLED [YUN 04, PAR 05, GHO 05, MEY 09, KEU 12, MAI 12]. The basic principle of ALD is to successively expose the surface where the mineral film is to be deposited on to successive chemical precursors to obtain ultrathin layers. Control reactions are carried out by the surface and not by the amount of reagent introduced because it is a self-limited growth process [PUU 05] (Figure 1.24).

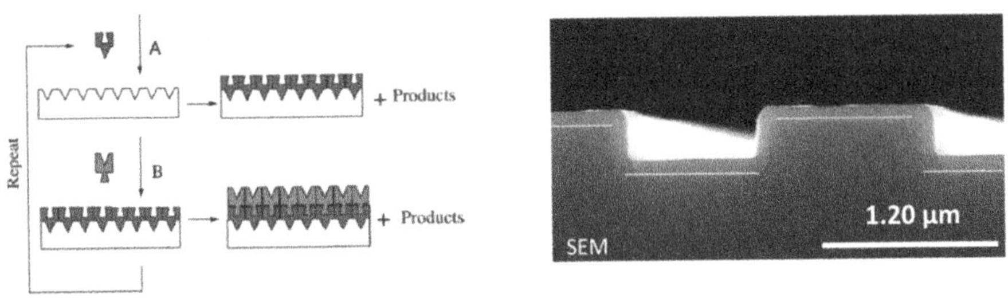

a) b)

Figure 1.24. *a) ALD principle (from Chem. Rev. 2010, 110, 111–131); b) demonstration of Al_2O_3 ALD conformality onto a Si trench (from CEA-LETI)*

It is common to assimilate ALD to the redox chemistry of trimethyl aluminum (TMA)/H_2O to make Al_2O_3. Indeed, this process is one of the simplest and most representative process of all processes in ALD technology. One of the main reasons lies in the general properties of the TMA molecule which is a highly volatile, reactive and at the same time temperature stable compound that shows a truly self-terminating reaction in ALD processing on different kinds of substrates. Its large temperature processing window between room temperature and ~300°C is really suitable for this material to be used as a barrier for encapsulating temperature sensitive devices like OLED, and especially OLED in microdisplays. Its fully thermal character turns out to be an asset compared to plasma-assisted processes like PECVD or plasma-enhanced ALD (PEALD) which may generate particles, detrimental to the goodness of the thin film encapsulation, in the gaseous phase [SEL 92].

It is worth noting here that particle contamination is a major concern when making OLED devices *a fortiori* microdisplays. Contamination can take place at every moment in the fabrication process. A clean room environment is mandatory for OLED mass-production. Indeed, a particle can be generated at different levels in the thin OLED architectures as described in Figure 1.25:

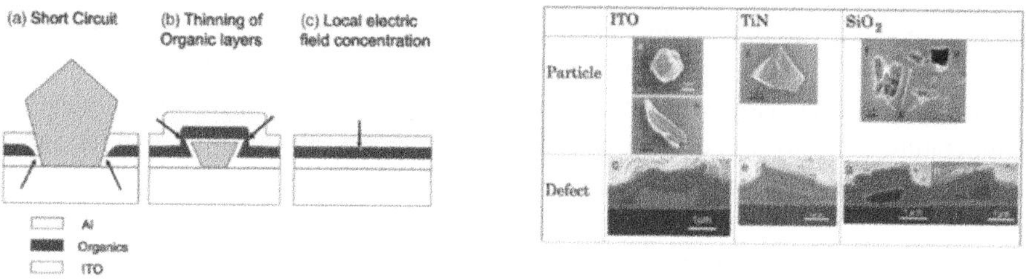

a) b)

Figure 1.25. *a) Schematic illustration of current leakage defects from particulate contaminants; b) SEM images of sputtered particles and resultant current leakage defects (source: [NAG 07])*

The different types of contamination yield different defects which turn into black spot defects which may or may not grow upon water ingress. It is obvious that for a microdisplay whose final goal is to be seen through a magnifying lens the presence of defects is detrimental to fabrication yield and, therefore, must be eradicated. As it seems rather difficult to cure a defect after the full OLED fabrication (OLED deposition + thin-film encapsulation + RGB color filter), we have to bear in mind that particle control is mandatory during the whole process. A recent EU project, Clean4Yield, led by Philips aims to propose solutions and strategies to decrease the particle levels in OLED fabrication [CLE 14].

Returning to ALD technology, a recent review has focused on probably the most pertinent publications of the last 10 years related to the use of Al2O3 ALD for barrier applications [PAR 11]. During this period, the outstanding quality of Al_2O_3 ALD films as barrier layers has been shown: it has been demonstrated that Al_2O_3 ALD films, deposited at low temperature (<120°C), can lower the H_2O permeability of plastic sheets to orders of magnitude below the permeability of the intrinsic sheet. Some WVTR values, as low as 1.7×10^{-5} g/m²/day, have been claimed for a 25 nm thick Al_2O_3 ALD oxide film [CAR 06]. The most striking asset for Al_2O_3 ALD barrier layers is their ability to achieve a high level of barrier quality onto polymer substrates with a critical thickness as low as 5 nm [CAR 09]. Intrinsic defects in thin solid films are usually related to film microstructure and the presence of pinholes. For organic electronics in general, thin inorganic films, whether they are electrodes of the device (a hypothetic defect-free Al cathode would be a high quality barrier by itself) or thin film barriers made by PVD or CVD derivative techniques, must be deposited at a low temperature (~100°C typically). As a result, their microstructure shows a certain level of porosity favored by columnar growth [THO 86]. Recently, Spee *et al.* have shown that a great difference exists between the WVTR of a PECVD-deposited SiN_x made at a temperature < 100°C and the WVTR of a PECVD deposited SiN_x made at 250°C [SPE 12]. By using a calcium-test experimental setup they concluded that the high temperature deposited barrier film presents far less defects and a denser microstructure which allow a calcium film to resist more than 30 days in 60°C/90% relative humidity (RH) conditions with minor degradation. However, the calcium film encapsulated with the single low temperature SiN_x layer became totally degraded after less than two days in the same testing conditions (Figure 1.26).

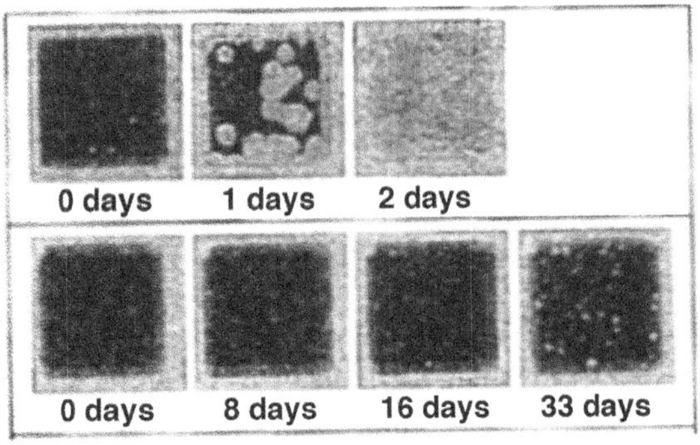

Figure 1.26. *Development in time of 5 x 5 mm² Ca squares covered with three different barrier layers and kept in a climate chamber at 60°C/90% RH; (top) low temperature SiNx, (middle) 100 nm state-of-art high temperature SiNx (source: Spee et al. [CSPE 12])*

Usually the density of pinholes in Al_2O_3 ALD films is supposed to be extremely low, almost non-existent [SUN 92]. Recently, Carcia *et al.* [CAR 10] demonstrated that permeation of water molecules in Al_2O_3 films occurs because of the presence of hydrogen atoms, bonded in the oxide structure as –OH, whose content could be as high as 10–20% for low temperature deposited Al_2O_3 ALD films. These hydroxyl groups create permeation pathways of high diffusivity for water molecules. In this formalism, the main assumption is that the film possesses no pinholes and that the permeation process rather follows a percolation model. Moreover, scientists have reported a defect density of 38/cm² for single Al_2O_3 ALD films (25 nm) deposited onto nickel substrates [ZHA 09]. Another publication from the same research group highlighted the dependency of the defect density of the oxide film on its thickness and the scientists found a way to decrease the defect density from $1.2 \times 10^{+5}$/cm² to 90/cm² for 2 nm thick Al_2O_3 ALD films just by introducing a thin ALD deposited W interlayer by ALD between the nickel substrate and the oxide film [ZHA 11]. Keuning *et al.* have recently reported a ~ 5/cm² density of defects for a 20 nm thick Al_2O_3 ALD film deposited by low temperature PEALD [KEU 12]. In this case, the authors assume that one dark spot on the OLED corresponds to one defect in the barrier layer. In this work, the authors have noticed that dark spots on the OLED emitting surface with the single Al_2O_3 layer (20 nm) show the highest growth rate upon time ($6 \times 10^{+3}$ µm²/h) compared to single thick layer (300 nm) of a-SiN_x:H deposited by PECVD (~$0.5 \times 10^{+3}$ µm²/h). Hence, the authors stated that the encapsulation mechanism for Al_2O_3 PEALD is mainly efficient for small defects present at the surface but that PEALD can leave larger defects, probably particles, either uncovered or the layer itself could suffer from local stress around the large defects. Another method presented recently uses a fluorescence test based upon the encapsulation of fluorescent AlQ_3 films [MAI 12]. This simple method allows visualizing defects in the encapsulation layers because AlQ_3 films become non-fluorescent if water penetrates the barrier down to the film. As a result, defects can be easily seen as black spots, as described in the following figure. Authors have been able to confirm the sensitivity of Al_2O_3 films to water ingress and their premature degradation in a hot humid environment. The main result was that Al_2O_3 is a good barrier that satisfies the specifications for OLED use but that its needs to be overprotected by an extra moisture stable layer like SiO_x [MAI 13].

This last method allows a counting of defects on the true OLED device and thus has the merit to be the most suitable test to provide if we want to evaluate the defects in ALD films used for OLED thin film encapsulation purposes.

Several methods exist to assess the WVTR of a thin-film barrier. Among them, the calcium-test has been already cited above but a typical permeameter measurement can also be performed. For that purpose, the barrier layer must be deposited on top of a semi-permeable membrane which does not turn out to be representative of the barrier properties deposited onto the real OLED architecture. For that reason we will not develop the detail of WVTR measurements in this book. Readers may instead find in the literature some reviews like the one proposed by Park *et al.* or Lewis and Weaver for OLED displays [PAR 11, LEW 04] or general WVTR measurements on the following web page: http://www1.eere.energy.gov/solar/pdfs/pvrw2010_dameron.pdf.

1.8. Specificities of OLED for microdisplays

OLED devices, which are now quite widely used for fabrication of small and medium-size OLED displays, should be necessarily modified or adapted to fit the specificities of OLED microdisplays.

There are some hurdles that must be overcome to guarantee fabricating a microdisplay using OLEDs. Most of them come from the fact that the microdisplay requires a monolithic integration of OLEDs on a CMOS active-matrix. For this aspect, the question is finally to know at what conditions OLED devices can be compatible with integrated circuits. The other difficulty comes from the fact that we want to achieve OLED pixel pitches in the order of 5–10 µm instead of 50 µm or more.

The specificities of making an OLED display using a CMOS circuit can be summarized as follows:

1) Voltage: CMOS circuits operate at low given voltages, for instance 3.3 V. Therefore, the OLED component must be highly efficient at very low voltage, compared with standard OLED devices used on small/medium size displays, which operate at ~ 7–8 V. This is achievable only by using electrically doped structures, or PIN OLED (see section 2.6).

2) Top-emission: the CMOS addressing circuit is made on silicon, which is totally opaque at visible wavelengths. Therefore, the only suitable structure for the OLED pixels is top-emitting. This implies a different choice for metal electrodes than those used for conventional bottom-emission displays. Also, in this case, cavity effects cannot be neglected and the pixel stack must thus be mastered at the nanometer scale to ensure good color reproducibility and a good efficiency. In mass production this may be a challenge if the OLED structure is too much complex, with too many layers to be reproduced.

3) Pixel topography: the CMOS circuit generally presents a surface topography that is not "naturally" suitable to OLED deposition. Furthermore, the last metal level of the circuit (foundry standards are Al, Al: Cu / TiN), which generally composes the anode of the OLED that is being deposited on, often shows a large surface roughness (>>1 nm) that can be severe for the reliability of the OLED, which is made up of very thin organic layers (~100 nm). Also, an electrically insulating resist

is deposited between the pixels creating a topography that can be several hundreds of nm thick.

4) Encapsulation: protection from oxidizing gases of atmosphere, by means of thin barrier films, of the OLED circuits deposited on the CMOS is probably the hardest task related to OLED integration on silicon-based backplanes. The encapsulation must first be transparent since the light comes out from the CMOS/top-emitting OLED pixels and must travel through the thin film barrier stack. But at the same time, the thin barrier film acts as a capping layer that modulates the transmittance of the upper semitransparent electrode (on which it has been deposited). In other words, a good compromise must be found between the barrier layer quality and the optimized optical extraction. Third, the encapsulation must allow for subsequent processes, for instance the realization of RGB color filters needed for microdisplays (direct photolithography of colored resins or lamination of RGB color filters made on glass wafers). All these conditions must also incorporate the topography of the circuit that requires encapsulation to be highly conformal at low deposition temperature (maximum allowed temperature for organic optoelectronics is <100°C).

5) The other difficulty in making OLED microdisplay is to achieve pixel pitches in the order of 5–10 μm instead of 50 μm or more. Regarding the active-matrix side, it is reasonably attainable, thanks to the very small feature size capability provided by the advanced patterning technologies of the CMOS manufacturing tools. The difficulty comes for achieving the color. As explained in section 2.3, conventional deposition method of OLED stacks is evaporation through a shadow mask. Therefore, to make color OLED displays, a first red stack is deposited through one mask, shielding the other colors (green and blue) pixel areas; then, the mask is subsequently moved to the next pixel to deposit the green, and then the blue stack. However, shadow mask technologies cannot be used for patterns smaller than typically 40–50 μm (see Chapter 6), so for OLED microdisplays another approach has to be found. This consists of depositing a full sheet of white emitting OLED, and obtaining color pixelization with color-filters. The realization of color filters can be achieved in two ways: on the one hand, direct deposition and photolithography, which allows high resolutions compatible with the pixel size of microdisplays (typ. < 10μm). On the other hand, color filters, previously processed on a glass wafer, can be laminated on the wafer containing the microdisplay chips. In the first case, this requires good encapsulation of OLED structure to withstand the three photolithography steps necessary to achieve the three RGB colors. In the second case, there will be a lighter constraint on the encapsulation, but the main hurdle is a need for ultraprecise hybridization tools to achieve alignment precision of less than 1 μm while having the smallest gap to avoid cross-talk between pixels.

The solutions found to address these difficulties for making an OLED microdisplay will be described in Chapter 6: OLED Microdisplay Fabrication.

1.9. Bibliography

[AFF 96] AFFINITO J.D., GROSS M.E., CORONADO C.A.., *et al.*, "A new method for fabricating transparent barrier layers", *Thin Solid Films*, vol. 290–291, p. 63, 1996.

[BÄS 81] BÄSSLER H., "Localized states and electronic transport in single component organic solids with diagonal disorder", *Physica Status Solidi (b)*, vol. 107, p. 9, 1981.

[BER 95] BERNANOSE A., "Electroluminescence of organic compounds", *British Journal of Applied Physics*, vol. 6, p. S54, 1995.

[BOR 91] BORSENBERGER P.M., BÄSSLER H., "Concerning the role of dipolar disorder on charge transport in molecularly doped polymers", *Journal of Chemical Physics*, vol. 95, p. 5327, 1991.

[BRÜ 13] BRÜTTING W., FRISCHEISEN J., SCHMIDT T.D., *et al.*, "Device efficiency of organic light-emitting diodes: Progress by improved light outcoupling", *Physica Status Solidi A*, vol. 210, p. 44, 2013.

[BUR 90] BURROUGHES J.H., BRADLEY D.D.C., BROWN A.R., *et al.*, "Light-emitting diodes based on conjugated polymers", *Nature,* vol. 347, p. 539, 1990.

[BUR 94] BURROWS P.E., FORREST S.R., "Electroluminescence from trap-limited current transport in vacuum deposited organic light-emitting devices", *Applied Physics Letters*, vol. 64, p. 2285, 1994.

[BUR 01] BURROWS P.E., GRAFF G.L., GROSS M.E., *et al.*, "Ultra barrier flexible substrates for flat panel displays", *Displays*, vol. 22, p. 65, 2001.

[CAR 09] CARCIA P.F., MCLEAN R.S., GRONER M.D., *et al.*, "Gas diffusion ultrabarriers on polymer substrates using Al_2O_3 atomic layer deposition and SiN plasma-enhanced chemical vapor deposition", *Journal of Applied Physics*, vol. 106, p. 023533, 2009.

[CAR 10] CARCIA P.F., MCLEAN R.S., REILLY M.H., "Permeation measurements and modeling of highly defective Al_2O_3 thin films grown by atomic layer deposition on polymers", *Applied Physics Letters*, vol. 97, p. 221901, 2010.

[CAR 06] CARCIA P.F., MCLEAN R.S., REILLY M.H., *et al.*, "Ca test of Al_2O_3 gas diffusion barriers grown by atomic layer deposition on polymers", *Applied Physics Letters*, vol. 89, p. 031915, 2006.

[CLE 14] HTTP://www.clean4yield.eu/. Accessed 2014.

[CON 56] CONWELL E.M., "Impurity Band Conduction in Germanium and Silicon", *Physical Review*, vol. 103, p. 51, 1956.

[CHA 96] CHATAM H., "Oxygen diffusion barrier properties of transparent oxide coatings on polymeric substrates", *Surface and Coatings Technology*, vol. 78, p. 1, 1996.

[FOR 00] FORREST S., BURROWS P., THOMPSON M., "The dawn of organic electronics", *IEEE Spectrum*, vol. 37, no. 8, p. 29, 2000.

[GIL 72] GILL W.D., "Drift mobilities in amorphous charge-transfer complexes of trinitrofluorenone and poly-n-vinylcarbazole", *Journal of Applied Physics*, vol. 43, p. 5033, 1972.

[GU 97] GU G., GARBUZOV D.Z., BURROWS P.E., *et al.*, "High-external-quantum-efficiency organic light-emitting devices", *Optics Letters*, vol. 22, p. 396, 1997.

[GEO 10] GEORGE S.M., "Atomic Layer Deposition: An Overview", *Chemical Reviews*. vol. 110, p. 111, 2010.

[GHO 05] GHOSH A.P., GERENSER L.J., JARMAN C.M., *et al.*, "Thin-film encapsulation of organic light-emitting devices", *Applied Physics Letters*, vol. 86, p. 223503, 2005.

[GRO 04] GRONER M.D., FABREGUETTE F.H., ELAM J.W., *et al.*, "Low-Temperature Al_2O_3 Atomic Layer Deposition", *Chemistry of Materials*, vol. 16, p. 639, 2004.

[GRO 06] GRONER M.D., GEORGE S.M., MCLEAN R.S., *et al.*, "Gas diffusion barriers on polymers using Al2O3 atomic layer deposition", *Applied Physics Letters*, vol. 88, p. 051907, 2006.

[HOF 10] HOFFMANN S., THOMSCHKE M., FREITAG P., *et al.*, "Top-emitting organic light-emitting diodes: Influence of cavity design", *Applied Physics Letters*, vol. 97, p. 253308, 2010.

[HUA 02] HUANG J., PFEIFFER M., WERNER A., *et al.*, "Low-voltage organic electroluminescent devices using pin structures", *Applied Physics Letters*, vol. 80, p. 139, 2002.

[KEU 12] KEUNING W., VAN DE WEIJER P., LIFKA H. *et al.*, "Cathode encapsulation of organic light-emitting diodes by atomic layer deposited Al_2O_3 films and Al_2O_3/a-SiNx:H stacks", *Journal of Vacuum Science and Technology A*, vol. 30, p. 01A131, 2012.

[KIM 07] KIM H.K., KIM M.S., KANG J.-W., *et al.*, "High-quality thin-film passivation by catalyzer-enhanced chemical vapor deposition for organic light-emitting diodes", *Applied Physics Letters*, vol. 90, p. 013502, 2007.

[LAN 06] LANGEREIS E.M., CREATORE M., HEIL S.B.S., "Plasma-assisted atomic layer deposition of Al_2O_3 moisture permeation barriers on polymers", *et al.*, *Applied Physics Letters*, vol. 89, p. 081915, 2006.

[LEW 04] LEWIS J.S., WEAVER M.S., "Thin-film permeation-barrier technology for flexible organic light-emitting devices", *IEEE Journal of Selected Topics Quantum Electronics*, vol. 10, pp. 45–57, 2004.

[LI 13] LI N., OIDA S., TULEVSKI G.S., *et al.*, "Efficient and bright organic light-emitting diodes on single-layer graphene electrodes", *Nature Communications*, vol. 4, no. 2294, p. 1, 2013.

[MAD 00] MADIGAN C.F., LU M.H., STRUM J.C., "Improvement of output coupling efficiency of organic light-emitting diodes by backside substrate modification", *Applied Physics Letters*, vol. 76, p. 1650, 2000.

[MAI 06] MAINDRON T., BEN KHALIFA M., VAUFREY D., *et al.*, "High performance and high stability PIN OLED", SID Symposium Digest of Technical Papers 37(1) 2002, 2006.

[MAI 12] MAINDRON T., SIMON J.Y., VIASNOFF E., *et al.*, "Stability of 8-hydroxyquinoline aluminum films encapsulated by a single Al_2O_3 barrier deposited by low temperature atomic layer deposition", *Thin Solid Films*, vol. 520, p. 6876, 2012.

[MAI 13] MAINDRON T., AVENTURIER B., GHAZOUANI A., *et al.*, "Investigation of Al_2O_3 barrier film properties made by atomic layer deposition onto fluorescent tris-(8-hydroxyquinoline) aluminium molecular films", *Thin Solid Films*, vol. 548, p. 517, 2013.

[MAR 62] MARK P., HELRICH W., "Space-Charge-Limited Currents in Organic Crystals", *Journal of. Applied Physics*, vol. 33, p. 205, 1962.

[MEY 09] MEYER J., SCHNEIDENBACH D., WINKLER T., *et al.*, "Reliable thin film encapsulation for organic light-emitting diodes grown by low-temperature atomic layer deposition", *Applied Physics Letters*, vol. 94, p. 233305, 2009.

[MIL 60] MILLER A., ABRAHAMS E., "Impurity Conduction at Low Concentrations", *Physical Review*, vol. 120, p. 745, 1960.

[NAG 07] NAGAI M., "Impact of Particulate Contaminants on the Current Leakage Defect in OLED Devices", *Journal of the Electrochemical Society*, vol. 154, p. J387, 2007.

[OSA 14] HTTP://www.osa-direct.com/osad-news/panasonic-develops-114-lmw-white-oled-for-lighting.html. Accessed 2014.

[PAR 05] PARK S.H.K., OH J., HWANG C.S., *et al.*, "Ultrathin Film Encapsulation of an OLED by ALD", *Electrochemical and Solid-State Letters*, vol. 8, p. H21, 2005.

[PAR 11] PARK J.-S., CHAE H., CHUNG H.K., *et al.*, "Thin film encapsulation for flexible AM-OLED: a review", *Semiconductor Science and Technology*, vol. 26, p. 034001, 2011.

[POP 63] POPE M., KALLMANN H.P., MAGNANTE P., *et al.*, "Electroluminescence in Organic Crystals", *Journal of Chemical Physics*, vol. 38, p. 2042, 1963.

[PUU 05] PUURUNEN R.L., "Surface chemistry of atomic layer deposition: A case study for the trimethylaluminum/water process", *Journal of Applied Physics*, vol. 97, p. 121301, 2005.

[REC 08] RECKZIEGEL S., KREYE D., PUEGNER T., *et al.*, "Optical sensors based on monolithic integrated organic light-emitting diodes (OLEDs)", *Proceedings of SPIE*, vol. 7003, p. 700316, 2008.

[ROB 79] ROBERTS G.G., MCGINNITY M., BARLOW W.A., *et al.*, "Electroluminescence, photoluminescence and electroabsorption of a lightly substituted anthracenelangmuir film", *Solid State Communications*, vol. 32, p. 683, 1979.

[SAV 97] SAVVATE'EV V.N., TARABIA M., CHAYET H., *et al.*, "Space-charge limited current and electroluminescence in conjugated polymer-based LEDs", *Synthetic Metals*, vol. 85, p. 1269, 1997.

[SEL 92] SELWYN G.S., PATTERSON E.F., "Plasma particulate contamination control. II. Self-cleaning tool design", *Journal of Vacuum Science and Technology A*, vol. 10, p. 1053, 1992.

[SOB 98] SOBRINHO A.S., CZEREMUSZKIN M.L., KLEMBERG SAPIEHA J.E., *et al.*, "Transparent barrier coatings on polyethylene terephthalate by single- and dual-frequency plasma-enhanced chemical vapor deposition", *Journal of Vacuum Science and Technology A*, vol. 16, p. 3190, 1998.

[SPE 12] SPEE D., VAN DER W.K., RATH W.J., *et al.*, "Excellent organic/inorganic transparent thin film moisture barrier entirely made by hot wire CVD at 100°C", *Physica Status Solidi Rapid Research Letters*, vol. 6, p. 151, 2012.

[SUN 92] SUNTOLA T., "Atomic layer epitaxy", *Thin Solid Films*, vol. 216, p. 84, 1992.

[TAN 87] TANG C.W., VAN SLYKE S.A., "Organic electroluminescent diodes", *Applied Physics Letters*, vol. 51, p. 913, 1987.

[THO 86] THORNTON J.A., "The microstructure of sputter-deposited coatings", *Journal of Vacuum Science and Technology A*, vol. 4, p. 3059, 1986.

[WIN 99] WINDISCH R., HEREMANS P., KNOLOCH A., *et al.*, "Light-emitting diodes with 31% external quantum efficiency by outcoupling of lateral waveguide modes", *Applied Physics Letters*, vol. 74, p. 2256, 1999.

[YAM 00] YAMASAKI T., SUMIOKA K., TSUTSUI T., "Organic light-emitting device with an ordered monolayer of silica microspheres as a scattering medium", *Applied Physics Letters*, vol. 76, p. 1243, 2000.

[YUN 04] YUN S.J., KO Y.-W., LIM J.W., "Passivation of organic light-emitting diodes with aluminum oxide thin films grown by plasma-enhanced atomic layer deposition", *Applied Physics Letters*, vol. 85, p. 4896, 2004.

[ZHA 09] ZHANG Y., BERTRAND J.A., YANG R., *et al.*, "Electroplating to visualize defects in Al_2O_3 thin films grown using atomic layer deposition", *Thin Solid Films*, vol. 517, p. 3269, 2009.

[ZHA 11] ZHANG Y., SEGHETE D., ABDULAGATOV A., *et al.*, "Investigation of the defect density in ultra-thin Al_2O_3 films grown using atomic layer deposition", *Surface and Coatings Technology*, vol. 205, p. 3334, 2011.

2

Overview of OLED Displays

As recalled in the brief history of organic electroluminescent devices in Chapter 1, it is only in 1987 that the first organic light-emitting diode (OLED) device involving all the baseline architecture of today's devices was proposed. Since then, it took only a decade to see the first commercial products, which are quite rare. Today, OLED devices are being widely spread around us, mainly (but not only) in display applications. In this chapter, we will review the applications addressed by OLED devices, by using a historical perspective, to illustrate how the scientific, technological and industrial improvements progressively provided to propose more of sophisticated products.

The first generation of OLED displays on the market was passive-matrix type. Then, small-size active-matrix OLED displays appeared. Later, there has been a continuous increase of resolution and size. This trend is expected to continue, along with new features such as transparency and flexibility. Finally, we will also review the new rising use of OLEDs: the lighting application.

2.1. Passive-matrix OLED displays

2.1.1. *Main characteristics*

The easiest way to display some information, is to use a group of single devices, arranged in such a way that they can represent simple characters such as alphanumeric signs (in general seven-segment device for numbers and letters) or icons. When it comes to making an *image*, we then need a *matrix* arrangement of dots, then making a matrix display. Two kinds of matrix addressing can be made: passive matrix or active matrix, the latter having additional active electronic devices to make the image.

Chapter written by François TEMPLIER.

PMOLED stands for passive-matrix OLED display. As it will be explained in more detail in Chapter 6, in a PMOLED display the image is written by sequential scanning of lines and applying the data signal to the columns. Since no switching device is present for each pixel, the pixels in each line are actually off most of the time. This means that all the desired luminance must be supplied within a very short time (actually the frame time divided by number of lines), which is obtained by supplying high voltages to the OLEDs. Such features put constraints on OLED lifetime and limit the number of lines and, behind this, display resolution. Being the easiest and cheapest way of making dot matrix OLED displays, PMOLED displays were the first commercially available and still constitute a consistent part of the market.

2.1.2. *Applications*

The first commercial application was a display for a car audio system as proposed by Pioneer in 1997. The model MEH-P9000 was a 5-in., multi-color (green/yellow/orange/blue) PMOLED display (Figure 2.1). It appeared to be an advantageous replacement of former-existing vacuum-fluorescent display (VFD), combining high resolution, high compactness and low consumption.

Figure 2.1. *First commercial application of OLED display for car audio system (from [PIO 14a]). This, and all other figures in this book, is available in full color at www.iste.co.uk/templier/OLED.zip*

Later, a wider range of applications for PMOLED emerged with MP3 players to display character data or small icons (Figure 2.2). More particularly for such small-sized devices, low power consumption and high compactness of OLED displays is appreciated or decisive. Also, with the rise and success of clamshell cell phones in the 2000s, PMOLED displays have been progressively and widely used. Initially, PMOLED displays were proposed only for the sub-display, where additional character data or icons can be displayed (for example: clock on external side, as shown on Figure 2.3). Such displays were monochrome, with a typical resolution of less than 100 × 100 pixels. Later still, PMOLED displays were used for the main display (Figure 2.4). For this purpose, it was then featuring color, but still with limited resolution.

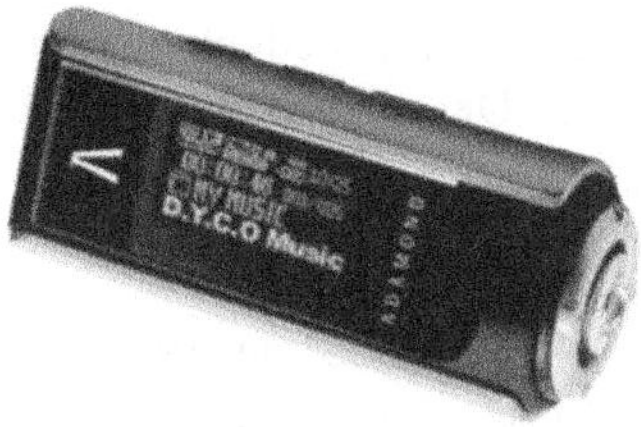

Figure 2.2. *Monochrome PMOLED for MP3 player. (Source: Adamond)*

Figure 2.3. *Monochrome 128 × 36 dots PMOLED for secondary display of clamshell cell phone. (Source: Sony-Ericsson)*

Figure 2.4. *1.9-In., 128 × 60 pixels, Color-PMOLED display, for main display of clamshell cell phone. (Source: Samsung)*

2.1.3. *Market and actors*

Today, PMOLED displays cover an even wider range of applications: car stereos cell phones, audio or video players and digital cameras.

The market for PMOLED displays was around $1 billion in 2013 and is not expected to grow in the next few years due to the need of higher resolution displays, which can only be fulfilled by active-matrix organic light-emitting diodes (AMOLEDs), unless new promising applications, such as smart watches, or instrument panels and gauges of the car, rise in the near future.

Today, PMOLEDs are being fabricated by several companies of which the largest are RiTdisplay, Pioneer, TDK and Visionox [OLE 14].

Pioneer

Pioneer, based in Japan, was founded in 1938, and produces home and car electronics systems. After being the first company to actually mass produce OLED displays in 1997, Pioneer currently proposes PMOLED displays with sizes ranging from 0.5 to 3.6-in. and resolution from 60 × 32 to 256 × 64; monochrome or color [PIO 14b].

TDK/ Futaba

TDK Corp. of Japan had a longstanding activity in OLED displays, but sold such activity to Futaba in 2011. Futaba proposes a wide range of PMOLED displays, with sizes ranging from 0.5 to 3.6-in. and resolution from 60 × 32 to 256 × RGB × 96. Display can be monochrome (white) or color (64 or 262 k); luminance is typically 300 and 150 cd/m² for white and color displays, respectfully [FUT 14]. Main features are shown in Figure 2.5.

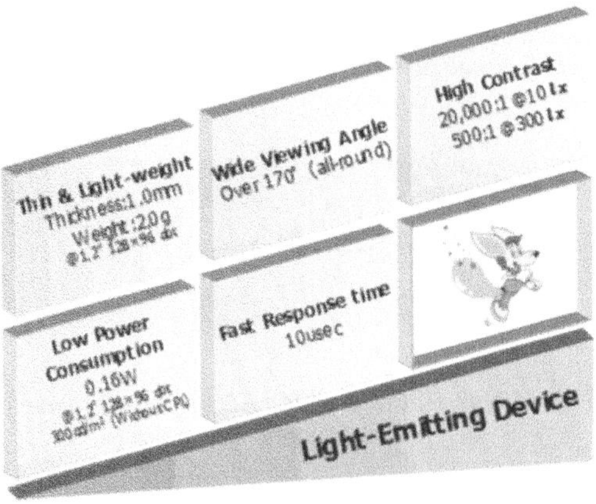

Figure 2.5. *Futuba OLED display main features (from [FUT 14])*

RitDisplay

Founded in 2000 in Taiwan, RitDisplay is a company, which was a spin off of RiTEK Corporation Flat Panel Display Division. RitDisplay proposes a wide range of PMOLED displays, with sizes ranging from 0.5 to 3.2-in. and resolution from 72 × 32 to 256 × 64. Display can be monochrome (white, yellow and blue) or color (64 or 262 k) (Figure 2.6).

Figure 2.6. *Some of RitDisplay products (from RIT 14])*

Applications of such displays are MP3 players, PDA, A/V displays, cellular phones, games, car stereos, dynamic information displays and digital cameras.

Winstar

Winstar, a Taiwan-based LCD, e-paper and PMOLED producer, is also developing transparent OLEDs and OLED lighting panels.

Visionox

Visionox is based in China and was founded in 2001. It is dedicated to developing, manufacturing and marketing display and lighting products, mainly OLEDs.

Visionox produces PMOLED panels of variable size from 0.9 to 2.8 in. (Figure 2.7), and was said to be the world's leading supplier in 2012.

Figure 2.7. *PMOLED from Visionox. Right: a monochrome, 2.8-in. with 256 × 64 pixels display (from [VIS 14])*

2.1.4. *Limitations/future of PMOLED*

Resolution limitation inherent to the passive addressing of PMOLEDs will limit its expansion as classical-type displays. However, one of the great features of OLED displays is that they can be made transparent and flexible. This will be illustrated in the further sections of this chapter.

2.2. Active-matrix AMOLED displays

2.2.1. *Main characteristics*

In active-matrix OLEDs, each pixel is connected to a transistor which is used as a switch. This device is turned on during the addressing phase and switched off to isolate the pixel (and maintain its state) while other pixels are being addressed. Also, at least another transistor is present in each pixel to be used as a current source for the OLED device, while optionally more transistors can be added for additional electronic functions such as threshold voltage shift compensation. Despite being more costly, active matrix provides fabricating OLED displays with a much greater number of pixels than PMOLED. Power consumption is reduced in the meantime. This advantage makes active-matrix OLEDs well-suited for portable electronics, in which power consumption is critical to battery life. The introduction of AMOLED to the market happened in 2006 with BenQ-Siemens, and rapidly grew in the following years mainly with the expansion of smartphones. The AMOLED market in 2013 was a huge at $10 billion, and growth is expected to continue.

2.2.2. *Applications: small and medium-size AMOLED*

Generally speaking, the display devices below 10 in. are defined as small and medium-sized display devices.

The first AMOLED display on the market was BenQ-Siemens S88 introduced in 2006, a 2.0-in. size with 176 × 220 pixels and 256K colors (Figure 2.8), followed by Nokia's 6215i. Later, size and resolution were increased, and in 2008 Nokia proposed the N85 cell phone featuring 2.6-in and 240 × 320 pixels (Figure 2.9).

Nokia was followed in 2010 by Motorola (Droid Razr), Samsung, LG and HTC (HTC Desire, HTC/Google Nexus One) both proposing cell phones with AMOLEDs and in 2012 by Sony (Xperia).

Figure 2.8. *First AMOLED display in market: 2.0-in. size with 176 × 220 pixels and 256K colors. (Source: BenQ-Siemens)*

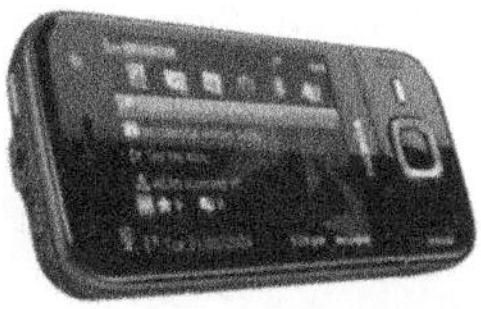

Figure 2.9. *2.6-in., quarter-VGA AMOLED display. (Source: Nokia)*

Samsung rapidly became the strong leader in smartphones with AMOLEDs, with the huge success of Galaxy S in 2010 featuring a 4.0-in., 800 × 480 pixels 'Super AMOLED' display (Figure 2.10). 'Super AMOLED' is Samsung's term for a display where the touch function is integrated in the display instead of being on top of it. Production shortage affected AMOLEDs in 2010 due to sudden high market growth, which required cell phone manufacturers, such as HTC, to propose LCDs on models, which were initially sold with AMOLEDs. This happened to Nexus One and HTC Desire models.

These AMOLED displays are competing with LCDs featured in high-end cell phones such as the Apple iPhone, HTC One X or Nexus 4. It is generally considered that OLEDs provide better contrast, whereas LCDs offer lower power consumption. In 2013, Samsung's Galaxy S4 presented a 5-in., full HD (1920 × 1080) AMOLED, which was competing with Apple iPhone 5 having 4-in. 1136 × 640 'Retina' LCD.

In the same year, the largest AMOLED on a cell phone was found on Samsung's Galaxy Note 3, with 5.7-in. size full HD. However, the biggest AMOLED display on mobile device was on Samsung Galaxy tab 7.7 with 7.7-in. and 1280 × 800 pixels (Figure 2.11).

Figure 2.10. *Samsung Galaxy S featuring 4.0-in., 800 × 480 pixels AMOLED display. (Source: Samsung)*

Figure 2.11. *Samsung Galaxy Tab 7.7 with 7.7-in., 1280 × 800 pixels AMOLED display (Source: Samsung)*

This product was the first computer to feature an AMOLED display, which might have marked the first step of the technology toward widespread computer and TV applications.

AMOLED is also present on some Samsung cameras, used as the backside display, with size around 3-in.

OLED displays in transitions

In summary, the small and medium size AMOLED display market has grown drastically in the years 2008–2013, pulled by a new generation of high-end cell phones, and more recently by new devices such as tablets. In 2013, active-matrix OLED display revenues were well above the $10 billion annual sales mark. That is still relatively small compared with the total market for LCD mobile devices, but it is significant from the standpoint of infrastructure investment, process development, and the materials and chemistry supply chain. It is also remarkable when we consider the outrageous dominance of LCD technology and its endless effort toward higher pixel density and ever-improving optical performance. It is hard to close the performance gap when the competing technology continues to improve in so many dimensions. Therefore, there is no doubt that the small and medium display market has been the battlefield where OLED technology has emerged as a new winning force.

2.2.3. *Applications: large-size OLED displays*

So far, AMOLED has not made a significant presence in the market of larger displays such as monitors or TVs. But, this trend might change rapidly.

2.2.3.1. *Monitors*

Since the late 2000s, Sony has proposed a range of AMOLED monitors for professional application. These are high-end products, featuring high performance such as hi-resolution, high contrast (black performance), quick response with virtually no motion blur, a wide color gamut and limited angle-dependence color shift. In 2013, size varied from 7.4, 17 and 25-in with resolution from Quarter HD resolution (960 × 540 pixels) to full HD (1920 × 1080) (Figure 2.12).

Figure 2.12. *Full HD (1920 × 1080), 25-in. AMOLED Professional Monitor. (Source: Sony)*

2.2.3.2. *Television*

In 2005, Samsung stunned the display world when it first showed a 40-in. prototype OLED TV (Figure 2.13). This breathtaking prototype demonstrated what everybody anticipated i.e. that OLEDs had several intrinsic performance advantages over LCDs owing to their nature as an emissive technology, including power consumption based on content, true black levels and very wide dynamic range, tunable color spectrums and very fast response times [KIM 09]. This also demonstrated that the game was on in terms of defining the future for OLED as a competing technology for large-format TVs as well. One of the events of 2007 in the world of display was the announcement by Sony to release the first commercial OLED TV, the XEL-1 (Figure 2.14). The XEL-1 was amazingly thin (3 mm), and had 960 × 540 resolution. It had an outstanding 1,000,000:1 contrast ratio, a terrestrial digital tuner, speakers, HDMI, USB and Ethernet jacks. It weighed 2 kg. Due to the very high price (more than €2000), this product was not really intended for mass production, and was discontinued a year later.

Figure 2.13. *Samsung 40-in. AMOLED TV prototype (2005). (Source: Samsung)*

Figure 2.14. *Sony XEL-1 AMOLED TV commercial product (2007). (Source: Sony)*

Also, in 2010 LG released for a short time an AMOLED TV, the EL9500 model, with an interesting 1366 × 768 resolution, but still at an expensive price (€ 1,700).

From 2013, new Tvs took a giant leap: Samsung (Figure 2.15) and LG (Figure 2.16) both proposing 55-in. models. The two models feature high definition and are actually curved.

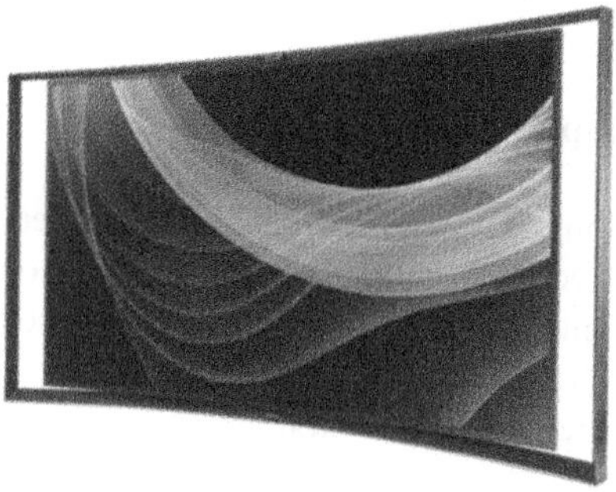

Figure 2.15. *Samsung KN55S9C, 55-in. curved AMOLED TV. (Source: Samsung)*

Figure 2.16. *LG 55EA980V, 55-in. curved AMOLED TV. (Source: LG)*

These TVs were then sold at around €8000 at the time of their launch (late 2013) which might be a limitation for significant sales. However, with such products, AMOLED TVs show that they are present in the core of the LCD TV business.

However, with LED-backlit 240-Hz panels featuring greatly improved image quality, contrast and wide viewing angles, in thicknesses less than 10 mm, the theoretical performance gap between TFT-LCDs and AMOLEDs continues to narrow. To achieve broad adoption of AMOLEDs in the TV market, cost reduction will be more important than performance improvements. Since AMOLEDs are simpler to manufacture than TFT-LCDs, having no backlight and color filter, and potentially no top glass, yielded costs could be only a fraction of that of TFT-LCDs. On the other hand, high unit costs for materials, low manufacturing yields and challenges in developing new manufacturing equipment and processes mean that even coming close to theoretical costs is quite difficult.

2.3. Trends in OLED displays: flexible and transparent

Flexible displays, along with transparent displays, certainly have much more assets than the 'wow' effect they generally produce on people. Indeed, they open a high potential of new applications and will certainly be serious growth drivers for matrix OLED displays. It is generally admitted that flexible display market will rapidly grow and reach a few billion dollars.

2.3.1. *Flexible and transparent PMOLED displays*

Making flexible PMOLED displays is potentially much easier than to make AMOLED, since they do not have active matrix. Such an electronic circuit, is made up of pattern of lines, columns and a few devices within each pixel (at least two transistors, one capacitor), and which is not straightforward to make flexible.

In 2012, Futaba unveiled a 3.5-in. 256 × 64 resolution, full-color flexible PMOLED prototype (Figure 2.17). The prototype was only 0.22 mm thick (compared to 2 mm of the regular rigid ones) and featured 100 cd/m^2 brightness. They then announced mass production to follow shortly, for both the consumer market, and for the automotive industry.

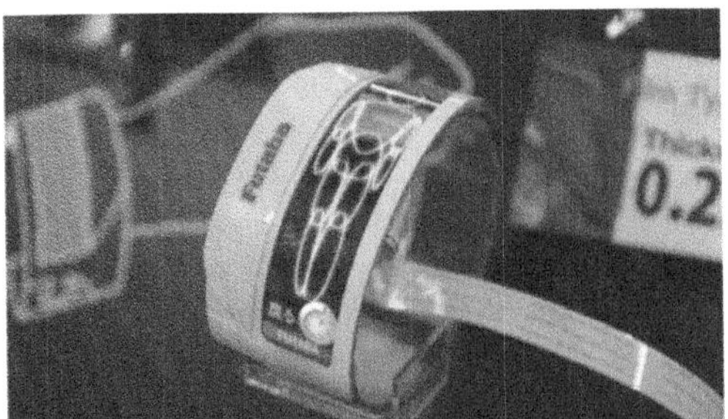

Figure 2.17. *3.5-in. 256 × 64 resolution, full-color flexible PMOLED prototype*

For the same reasons as for the flexibility, making transparent PMOLED displays is somewhat easier than AMOLEDs, since there are no opaque patterns shielding the light.

In May 2011, TDK announced they would be starting mass production of transparent PMOLED displays for mobile phones and other applications. TDK's transparent panels offered 320 × 240 resolution with 50% transmittance and were 2" in size (Figure 2.18). The same year, Lenovo announced the S-800 cellphone, featuring a transparent, monochrome, 2.4-in., 320 × 240 (QVGA) resolution PMOLED display.

Figure 2.18. *2-in. transparent monochrome, 320 × 240 resolution PMOLED with 50% transparency. (Source: TDK)*

OSD Display, a US company, also offers transparent PMOLED displays, either small monochrome dot-matrix displays (0.56"–1.54" active area) with a transmittance of 70% or segmented displays which are larger (up to 3") and with a transmittance of 45% (Figure 2.19).

Figure 2.19. *Segmented large monochrome OLED display with 45% transparency. (Source: OSD)*

2.3.2. *Flexible AMOLED displays*

Though many fully flexible AMOLEDs have been shown as prototypes in the last 10 years, the first commercial products are now arriving on the market in an intermediate state: namely curved displays, exhibiting a fixed, high-radius curvature.

In October 2013, Samsung announced the Galaxy Round, claiming it to be the first curved OLED phone (Figure 2.20). The handset features a curved 5.7 in. full HD (1920 × 1080). The display is curved in the horizontal direction, with a radius of 400 mm. The advantage of this feature is not indicated.

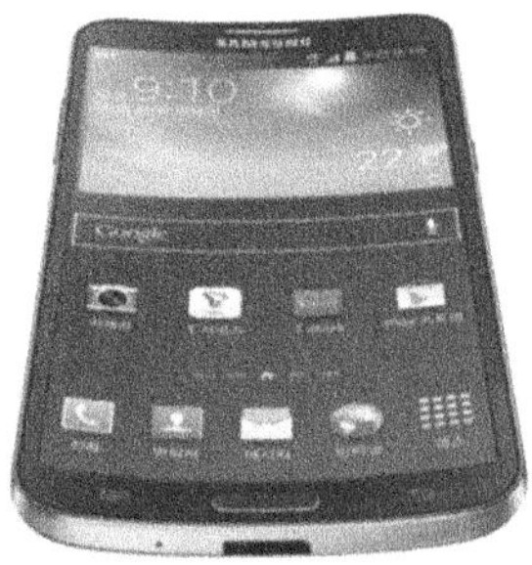

Figure 2.20. *5.7-inch full HD (1920 × 1080) curved (radius: 400 mm) AMOLED display. (Source: Samsung)*

Figure 2.21. *6.0-in. 720 × 1280 pixels curved (radius: 700 mm) AMOLED display. (Source: LG)*

At the same time, LG announced they would start mass production of flexible OLED smartphone screens, and shortly after the new LG G Flex smartphone featuring a curved 6.0 in. 720 × 1280 pixels AMOLED (Figure 2.21). Unlike the Samsung phone, curvature is in the vertical direction, with a radius of 700 mm.

Actual flexible AMOLED displays should arrive later.

Transparent AMOLED display prototypes have already been shown, but no commercial product has been released so far.

2.4. OLED lighting

OLED technology became rapidly attractive for lighting applications, for at least two reasons:

– interesting luminous efficiencies: lighting devices equipped with the high efficiency OLEDs are considered to achieve a luminous efficiency of 30–60 lm/W, much higher than bulb or halogen counterparts, close to that of fluorescent lamps (approximately 60–80 lm/W) or LED lighting fixtures (approximately 80 lm/W) currently available on the market;

– OLEDs are deposited as thin films, therefore, they provide unique feature of *surface* lighting, which gives an almost unlimited freedom to designers for imagining and making any kind of shape, even with curved surfaces, and with the possibility of choosing (and even tuning) the color.

Unlike conventional lighting sources, OLED devices will have added value when used as additional or ambient lighting rather than the main/powerfull light source.

OLED light sources are commercially available from companies such as Philips (Figure 2.22), OSRAM (Figure 2.23) and Panasonic.

Panasonic announced commercialization of OLED panels in 2011. The product is a light source with a panel section of 97 × 90 mm, as thin as approximately 2 mm, featuring high luminance (3,000 cd/m^2) as well as 30 lm/W luminous efficiency, 10,000 h durability (with 70% lumen maintenance factor), and color temperature of 3,000, 4,000 or 5,000 K [PAN 14].

Figure 2.22. *OLED light panels from Philips (ref GL350), size 124,5 × 124,5 mm, brightness of 4,000 cd/m^2 and efficiency of 16.7 lm/W. (Source: Philips)*

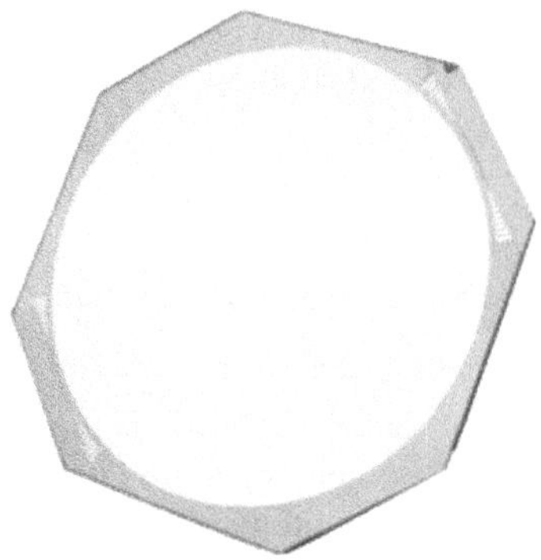

Figure 2.23. *OLED light panels from OSRAM (ref ORBEOS CDW-031), size 90.0 mm width and 1,000 cd/cm² brightness, and efficiency of 23 lm/W. (Source: OSRAM)*

The remaining challenges for OLED lighting are the following:

– Lifetime: despite tremendous progress in the past few years, lifetime should be improved, particularly to compete with other lighting techniques such as LEDs. LED lamps regularly offer in excess of 50,000 h, which is why they initially found a niche market in out-of-reach outdoor applications. In contrast, OLED lighting offers 5,000–15,000 h of operational life even when encapsulated.

– Cost: they still cost \$300–\$500/klm at panel level, which is still too high compared to the present low cost of LED lighting, selling at \$5/klm at package level. This will be the main issue to solve for a high-volume entry onto the market.

2.5. Microdisplays

OLED microdisplays are the main subject of this book. As stated in Chapter 1, they were first proposed on the market in 2001 by eMagin. Since then, other manufacturers have emerged, and they will be listed in a dedicated section of Chapter 9. But, one thing is important to mention here: the first conventional OLED displays appeared in 2003 (Kodak PMOLED on camera). Therefore, the first OLED microdisplays were on the market *before* their conventional medium or large-size counterpart! This is opposite to the case of LCD displays where microdisplay versions appeared much later the conventional ones.

2.6. Bibliography

[FUT 14] HTTP://www.futaba.co.jp/en/display/oled/index.html. Accessed 2014.

[OLE 14] HTTP://www.oled-info.com/companies-list/oled-display-suppliers. Accessed 2014.

[PAN 14] HTTP://panasonic.co.jp/es/pesiol/en/product/index.html. Accessed 2014.

[PIO 14a] HTTP://pioneer.jp/press-e/index/1484. Accessed 2014.

[PIO 14b] HTTP://pioneer.jp/topec/jigyo/oled/el_seisan-e.html. Accessed 2014.

[RIT 14] HTTP://www.ritek.com/p2-pro3-oled.asp. Accessed 2014.

[VIS 14] HTTP://www.visionox.com/en/color.aspx. Accessed 2014.

[KIM 09] KIM H.D., KIM H.J., BERKELEY B.H., *et al.*, "Emerging technologies for the commercialization of AMOLED TVs", *Information Display*, vol. 25, no. 9, September 2009.

3

OLED Characterization

The efficient conversion of electrical to radiated power in organic light-emitting diodes (OLEDs) is one of the primary goals in the development of these devices for technological application such as display.

But to achieve high efficiency and large gamut OLEDs, prior to OLED fabrication, first requires focusing on organic semiconductor properties to screen materials. So, the first part of this chapter deals with organic semiconductor characterization. Then, the second part is dedicated to OLED device characterization.

In this chapter, organic semiconductor physical properties, such as glass transition temperature [CON 94, KIM 09], purity [NOR 04] and granulometry will not be discussed.

3.1. Electronic properties of organic semiconductors

It is well recognized that facile injection of electrons at the cathode and holes at the anode is a desirable characteristic of the OLED structure, and experimental efforts have focused on finding suitable materials of low and high work function to achieve ohmic contact with electron transporting material and hole transporting material, respectively.

Consequently, the first requirements for optimizing OLED efficiency are to determine the electronic properties such as highest occupied molecular orbiltal (HOMO) and lowest unoccupied molecular orbital (LUMO) energy levels of organic materials.

Chapter written by David VAUFREY.

3.1.1. *HOMO and LUMO level determination*

3.1.1.1. *X-ray photoelectron spectroscopy and ultraviolet photoelectron spectroscopy*

X-ray photoelectron spectroscopy (XPS) is a surface-sensitive quantitative spectroscopic technique. XPS spectra are obtained by irradiating a material with a beam of X-rays while simultaneously measuring the kinetic energy and number of electrons that escape from the top 0 to 10 nm of the material being analyzed. XPS requires high vacuum ($P \sim 10^{-8}$ millibar) or ultra-high vacuum (UHV; $P < 10^{-9}$ millibar) conditions.

Ultraviolet photoelectron spectroscopy (UPS) refers to the measurement of kinetic energy spectra of photoelectrons emitted by molecules, which have absorbed ultraviolet photons, in order to determine molecular orbital energies in the valence region.

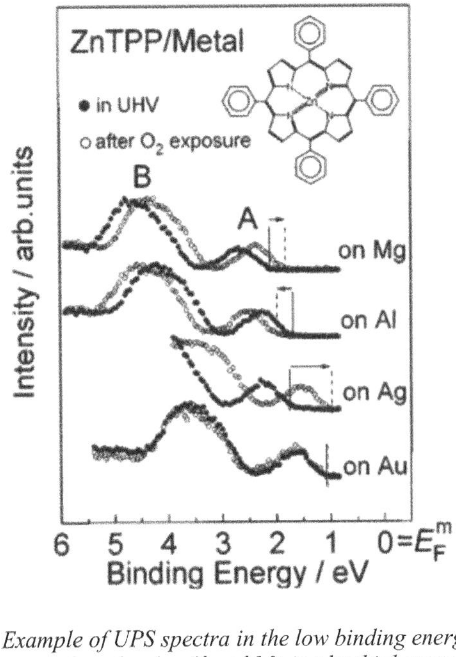

Figure 3.1. *Example of UPS spectra in the low binding energy region for ZnTPP films evaporated on Au, Ag, Al and Mg in ultrahigh vacuum (filled circles) and after exposure to O_2 (open circles) [NAR 95]. This figure, and all other figures in this book, is available in full color at www.iste.co.uk/templier/OLED.zip*

In the present example of Figure 3.1, similarity of UPS spectral shape among the spectra indicates the absence of change in the molecular electronic structure of ZnTPP. The peak position depends significantly on the metal, and is a linear function of metal work function. So, the peak position is proportional to the barrier height between the metal work function and the energy of the top of the valence state of ZnTPP.

3.1.1.2. *Cyclic voltammetry*

Cyclic voltammetry (CV) is a type of potentiodynamic electrochemical measurement. In a cyclic voltammetry experiment, the working electrode potential is ramped linearly versus time, like linear sweep voltammetry. When cyclic voltammetry reaches a set potential, the working electrode's potential ramp is inverted. This inversion can happen multiple times during a single experiment. The current at the working electrode is plotted versus the applied voltage to give the cyclic voltammogram trace.

CV and differential pulse voltammetry (DPV) are useful and valuable tools to study reversible redox behavior, electrochemical stability and to get information about HOMO and LUMO energy values of hole and electron transport materials and to examination of the barriers for charge injection.

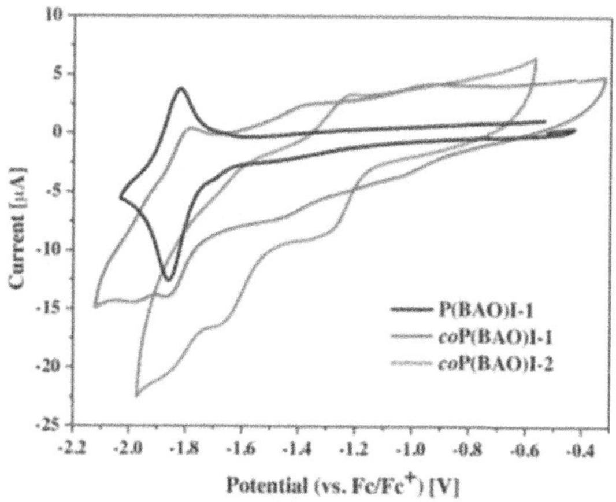

Figure 3.2. *Cyclic voltammograms example (CV) of P(BAO)I-1, coP(BAO)I-1 and coP(BAO)I-2 during electrochemical reduction at scan rate 100 mV/s [GRU 14]*

Gap, HOMO and LUMO levels can be deduced from oxidation and reduction potentials.

3.1.1.3. *Kelvin probe microscopy*

Kelvin probe force microscopy (KPFM), also known as surface potential microscopy, is a non-contact variant of atomic force microscopy (AFM), with which the *work function* of surfaces can be observed at *atomic* or *molecular* scales.

Recent works report band bending measurements in organic semiconductors using the Kelvin probe method [ISH 04].

To the extent that good injection results in space-charge limited current flow in organic semiconductors, it is then clear that higher carrier mobility leads to lower OLED operating voltages. Consequently, to increase OLED quantum efficiency, organic semiconductor mobility must be known.

3.1.2. *Mobility measurement*

Not all these methods, detailed hereafter, are suitable for a particular sample and the choice of the method depends on the details, such as sample thickness, nature of the electrodes and the energy levels of the materials.

3.1.2.1. *Space charge limited current: dark injection SCLC and trap free*

The sample utilized for characterization of charge transport by dark injection (DI) space-charged limited current (SCLC) technique normally consists of a thin organic semiconductor (~200 nm^{-2} μm in thickness) sandwiched between two electrodes. This technique requires one charge injection contact and one injection-blocking contact. Theoretical calculation and experimental evidence indicate that the injection barrier should not exceed 0.3 eV for SCLC experiments [MAL 99, LIU 07, DAV 97]. In DI, a step function voltage is applied to the injection contact of the sandwiched sample. The typical observed current transient (Figure 3.2) shows a peak at τ_{DI}, which relates to the free carrier transit time t_{tr} by [LAM 70]:

$$\tau\mathrm{DI} = 0.787\,ttr$$

The t_{tr} calculated using the above given equation can then be utilized to determine the charge carrier mobility through [SHI 01]:

$$\mu = \frac{v}{E} = \frac{d/t_{tr}}{V/_{d}} = \frac{d^2}{Vt_{tr}}$$

where V is the step function voltage, d is the sample thickness. An increase in the current to a maximum value is observed as the leading front of the charge carriers is transported through the organic semiconductor film. Under the applied field, the leading front of space charge will advance through the film with time. As the charges advance through the film, the gap between the counter electrode and the leading front decreases and causes an increase in the electric field. This results in a faster transit time, compared with the space charge-free transit time. Consequently, the current density at the peak is slightly higher than the equilibrium SCLC density. For longer duration, the charge distribution reaches a dynamic equilibrium and the equilibrium SCLC density is reached. Dramatically, different current transients can be obtained for DI SCLC measurements with different electrical contacts; hence, an ohmic contact is necessary for these measurements. DI SCLC experiments also require a much smaller resistor condensator (RC) time constant of the utilized experimental circuit when compared with the measured τ_{DI} so that the τ_{DI} can be correctly determined. If the obtained τ_{DI} is close to the RC time constant, it is very possible that the RC effect has significant contribution on τ_{DI}, and the RC time constant needs to be reduced until RC < τ_{DI}. Temperature and field dependence of the mobility can be obtained using the DI SCLC technique. This method has been successfully used to measure hole mobility for small molecule organic semiconductors [POP 03, SCO 99, CHI 11] and hole mobilities of polyfluorene copolymers [CAM 00, CAM 01, ESW 11].

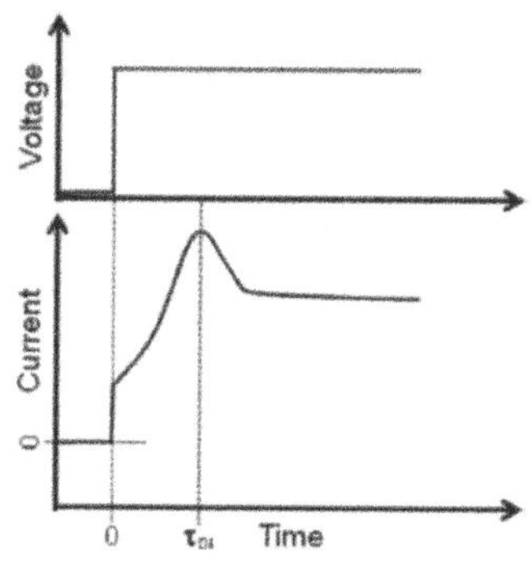

Figure 3.3. *A representative schematic of an ideal DI current transient*

The trap-free (TF) SCLC technique requires similar sample geometry and electrode Fermi levels as described above in the DI SCLC technique [POP 03]. For the TF case with both ohmic contacts and ignoring surface potential jumps, the *J–V* relation is given by the Mott–Gurney equation [MOT 40]:

$$J = \frac{9}{8}\mu\varepsilon\varepsilon_0\frac{V^2}{d^3}$$

where *J* is the current density, *V* is the applied voltage, 1 is the mobility, ε is the dielectric constant, and ε_0 is the vacuum permittivity. For materials that display field dependence of mobility, often the equation is modified to [MUR 70, MAT 09, TAN 04],

$$J = \frac{9}{8}\varepsilon\varepsilon_0\mu_0 e^{0.89\gamma\left(\frac{V}{d}\right)^{1/2}}\frac{V^2}{d^3}$$

where μ_0 is the zero field mobility and γ describes the field activation, that is, the field dependence of mobility as a function of temperature. Using the time of flight (TOF) technique (discussed in a following paragraph) on molecularly doped polyvinylcarbazoles, Gill proposed the temperature dependence of γ using the following equation [GIL 72]:

$$\gamma = \mathrm{B}\left(\frac{1}{\mathrm{k_B T}} - \frac{1}{\mathrm{k_B T_0}}\right)$$

where kB is the Boltzmann constant, $B = 2.9 \times 10^{-5}$ eV $(m/V)^{1/2}$ and $T_0 = 600$ K. The values for B and T_0 are similar for a large class of disordered organic semiconductors and that the variation in parameters describing this behavior is less than a factor of two [BOR 93]. Using the temperature dependence of dc mobility with applied field as a parameter plot [MAR 99] and the mobility as a function of $E^{1/2}$ plot, these constants can also be independently determined.

As the required thickness of the sample is similar to that used for the active layer in bulk heterojunction solar cells, many researchers use this technique for charge-carrier mobility measurements [LI 11]. However, a charge injection contact is critical for SCLC.

3.1.2.2. *Time of flight*

Time of flight (TOF) is a widely used technique for determination of charge carrier mobilities in organic semiconductors [SPE 69]. Similar sample geometry is utilized for the TOF samples as stated above, with the exception of the charge injection layer. In addition, much thicker films (>1 μm) of the organic semiconductor are required. As photogenerated charge carriers are required in TOF, transparent or semitransparent electrodes are also necessary. Furthermore, injection-blocking contacts are required for this technique. If this condition is not fulfilled, the injection of carriers occurs under the strong applied electric fields leading to SCLC, which can significantly affect the applied electric field, making the electric field extremely non uniform. Typically, in TOF measurements, a thin sheet of charge carriers is created in the organic semiconductor using a short laser pulse incident on the sample through the transparent/semitransparent electrode. Depending on the polarity of the applied uniform DC electric field (E) either electrons or holes are transported across the sample. In general, before utilizing the TOF technique, following parameters should be considered:

dielectric relaxation time (τ_σ) should be larger than the transit time (t_{tr}), sample thickness should be significantly larger than the inverse of the absorption coefficient (i.e. d > 1/α), the RC time constant should be much smaller than t_{tr} (differential TOF technique) and the integral TOF technique should be used when RC time constant is much larger than t_{tr}. In the most commonly used low light intensity mode of the TOF technique, the absorption depth of the utilized laser pulse should be short enough such that the photogenerated carriers are concentrated in a thin layer with well-defined position, compared to the sample thickness *d*. This requirement can be written as:

$$d \gg \frac{\ln 10}{\alpha}$$

where α is the absorption coefficient of the material. About 90% of the photo generated carriers are concentrated in a thin layer with the width of ln10/α. The requirement can also be expressed as:

$$A_\lambda = \frac{\alpha d}{\ln 10} \gg 1$$

where A_λ is the absorbance (or optical density) of the sample at the excitation laser wavelength λ. Under the presence of the electric field, the thin layer of charge carriers travels from the front electrode to the back electrode. The duration of the light pulse should be short as compared to the transit time of the charge carriers. As the sheet of charges reaches the back electrode, a drop in the photocurrent is observed and a characteristic plateau and ''knee'' at the transit time t_{tr} is observed in the photocurrent transient, especially for materials displaying non dispersive charge transport. The charge transport in disordered organic semiconductors can also be dispersive in nature. This is due to the existence of a distribution of charge trapping sites with different depths. In the photocurrent transient, the characteristic plateau region is not observed for materials displaying dispersive charge transport. In this case, the photocurrent transient is plotted in a double-logarithmic format and the t_{tr} is determined from the inflection point [SCH 75]. The charge carrier mobility can be determined using the following equation:

$$\mu = \frac{d^2}{t_{tr} V}$$

where *V* is the applied voltage. If the organic semiconductor does not have enough absorption or thickness to satisfy the condition expressed in equation, a charge generation layer with strong absorption at the laser wavelength can be added between the sample and the front electrode.

An important requirement of the TOF technique is that the transit time t_{tr} should be shorter than the dielectric relaxation time τ_{σ}, which is defined as the time for the photogenerated charges to relax back to the original state [LAM 70]. The dielectric relaxation time is defined as,

$$\tau_{\sigma} = \frac{\varepsilon\varepsilon_0}{\sigma}$$

where σ is the conductivity of the material. In TOF measurements, the length of the region where charge depletion occurs should be greater than the sample thickness [JUS 02]. Significant overestimation of the true mobility can occur if this condition is not satisfied, leading to a false-negative field dependence at low field.

An advantage of the TOF technique is that mobility of both electron and hole can be obtained from the same sample [CHE 00, WON 07].

3.1.2.3. *Carrier extraction by linearly increasing voltage and photo-CELIV*

Carrier extraction by linearly increasing voltage (CELIV) is a mobility measurement technique, which has been applied more recently to the characterization of charge transport in organic semiconductors [JUS 00a]. The sample geometry and the required energy levels are similar to the one used in TOF technique. The experimental setup for CELIV is similar to that utilized in TOF, except that a linearly increasing voltage (ramp) pulse with slope, $A = U/t_{pulse}$ (where t_{pulse} is the duration of the ramp), is applied to the sample [JUS 00a]. A typical current transient from a CELIV measurement is also displayed in Figure 3.4.

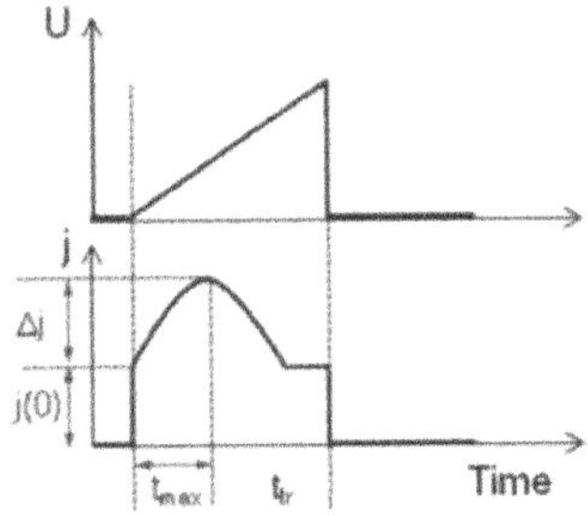

Figure 3.4. *A representative schematic for an ideal current transient obtained in CELIV for the applied voltage ramp*

The increase of initial step in the current is due to the capacitance of the sample, given by:

$$j(0) = \frac{\varepsilon\varepsilon_0 A}{d}$$

where d is the film thickness [CHE 00]. The increase in current after $j(0)$ is caused by the extraction of the dominant charge carrier present in the sample. The time required to reach the maximum in the extraction current (t_{max}) is utilized for calculating the charge carrier mobility. The mobility is obtained from measured current by the following expression:

$$\mu = \frac{2d^2}{3At^2_{max}}\left[1 + 0.36\frac{\Delta j}{j(0)}\right] - 1$$

where A is the rate of increase of the applied voltage, Δj is the difference between the maximum current and $j(0)$, as shown in Figure 3.9. The above equation is valid when $\Delta j/j(0)$ is less than or close to one. In the case where $\Delta j/j(0) < 7$, numerical solutions to the equations describing the current transient yields a more precise expression for mobility given by [BAN 10]:

$$\mu = \frac{2d^2}{3At^2_{max}}\left[0.329e^{-0.180\frac{\Delta j}{j(0)}} + 0.005e^{0.253\frac{\Delta j}{j(0)}}\right]$$

Typically, in organic semiconductors, the equilibrium charge carriers are present due to low-level doping or impurities. In the case of materials that do not contain these equilibrium charge carriers, irradiation of sample with a short laser pulse can be utilized to obtain photogenerated charge carriers, which after a certain delay can then be extracted from the film [GEN 03, MOZ 05].

3.1.2.4. *Field effect transistor*

The source and the drain electrodes in field effect transfers (FETs) form an ohmic contact with the conducting channel. The source electrode is grounded and all other voltages as given in relation to this electrode.

The charge carrier mobility is determined using the characteristic output curves of the transistor, that is, the plot of drain–source current (I_{DS}) versus drain voltage (V_{DS}) as a function of varying gate voltages (V_{GS}).

The output curves can be differentiated into two regions: the linear region and the saturation region. The linear region of the curve is defined as:

$$I_{DS} = \frac{W}{L}\mu C_S\left((V_{GS} - V_T) - \frac{V_{DS}}{2}\right)V_{DS}$$

where W and L are the width and length of the conduction channel, respectively; μ is the charge carrier mobility, CS is the capacitance per unit area for the insulating layer and V_T is the threshold voltage. The slope of the transfer characteristic curve, which is a plot of I_{DS} versus V_{GS} at a constant V_{DS}, can be utilized to determine the charge carrier mobility using the following equation:

$$\frac{dI_{DS}}{dV_{GS}} = \frac{W}{L}\mu C_S V_{DS}$$

However, the above given equation can be utilized only when $V_{DS} < (V_{GS} - V_T)$. The saturation region ($V_{DS} > V_{GS} - V_T$) of the output curve is defined as:

$$I_{DS} = \frac{W}{2L}\mu C_S(V_{GS} - V_T)^2$$

For the saturation region, the slope of the transfer characteristic (I_{DS}^2 vs. V_{GS} at constant V_{DS}) can also be utilized for calculating the charge carrier mobility. The above given equations are valid in the case where the charge carrier mobility is assumed to be temperature and field independent. But most organic semiconductors display field as well as temperature-dependant charge carrier mobilities [HOR 99, HOR 10, JUS 00b, VER 04].

3.2. Optical properties of organic semiconductors

3.2.1. *Spectrometry*

It is essential to know the optical constants of organic semiconductors, which compose the OLED stack, since refractive index n and absorption index k could impact the OLED optical out coupling.

The measurement of the $T(\lambda)$ spectra and reflectance spectra $R(\lambda)$ for the organic semiconductor at normal incidence light in the visible wavelength range could confirm or infirm that organic semiconductors are transparent and no light is scattered or absorbed (transparent region, $T + R = 1$ and absorbing region $R + T < 1$). The refractive index $n(\lambda)$ and the absorption index $k(\lambda)$ could be computed using the obtained data from transmittance and reflectance spectra. The procedure consists of considering that the normal reflectance for any absorbing material can be given by the following equation [CAG 09]:

$$R = \left(\frac{(n-1)^2 + k^2}{(n+1)^2 + k^2}\right)$$

Then, the refractive index can be computed directly from the solution of previous equation [YAK 07]:

$$n = \left(\frac{1+R}{1-R}\right) + \left(\frac{4R}{(1-R)^2} - k^2\right)^{1/2}$$

The values of the absorption index k can be calculated from the absorption coefficient using the relation $k = \alpha\lambda/4\pi$ [SWA 83], where α = A/t, and A is the measured absorbance and t is the thickness of the thin film.

The optical properties of any solid material are characterized by the complex refractive index n = n(λ)+ik(λ) and the dielectric function $\varepsilon = \varepsilon_1(\lambda) + i\varepsilon_2(\lambda)$. The real part ε_1 (represents the normal dielectric constant) and the imaginary part ε_2 (represents the absorption associated with radiation by free carrier) of the dielectric constants were determined by the following relations in the absorption region ($k \neq 0$) [FAD 09].

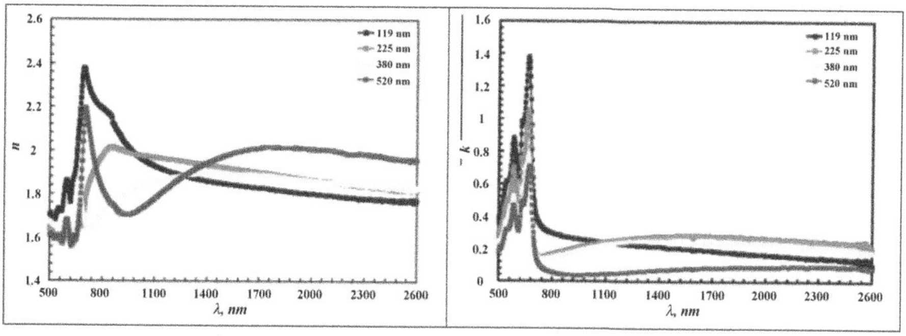

a) b)

Figure 3.5. *a) Example of dependence of refractive index and b) absorption index on the wavelength for pentacene thin films of different thicknesses*

$$\varepsilon_1 = n^2 - k^2 = \varepsilon_\infty - \left(\frac{e^2N}{4\pi\varepsilon_0 m * C^2}\right)\lambda^2$$

and

$$\varepsilon_2 = 2nk = \left(\frac{\varepsilon_\infty \omega_p^2}{8\pi^2 C^3 \tau}\right)\lambda^3$$

where τ is the optical relaxation time and $\omega_p = 2\pi\upsilon$ is the photon angular frequency. The variation of the dielectric constants with the photon energy indicates that there are some interactions between photons and electrons in the organic semiconductors. These interactions are observed in the shapes of the real and imaginary parts of the dielectric function and they are the cause of the formation of peaks in the dielectric spectra which depends on the material type [FAD 09]. The spectral distribution of the absorption coefficient, for a semiconductor, near the fundamental edge is very helpful when investigating the type of the optical transitions taking place in a material (direct or indirect interband transitions) as well as for its energy gap. Therefore, the optical band gap and nature of the optical transitions can be obtained from the dependence of the absorption coefficient on the photon energy ($h\upsilon$). In many organic materials, the absorption coefficient can be analyzed at the fundamental edge in terms of band-to-band transitions theory [ELN 04]. In that treatment, the absorption data follow a power-law behavior that is given by [PAL 93]:

$$\alpha\, h\upsilon = A(h\upsilon - E_g)^m$$

where α is the absorbance coefficient, $h\upsilon$ is the photon energy, E_g is the optical band gap, A is a constant that is independent of the photon energy, but depends on the transition probability. m is a number, which characterizes the transition process. In this chapter, the indirect band gap is determined by plotting $(\alpha h\upsilon)^{1/2}$ versus the photon energy ($h\upsilon$). With extrapolation of the linear part of the curve to lower energy, it is possible to determine the band gap.

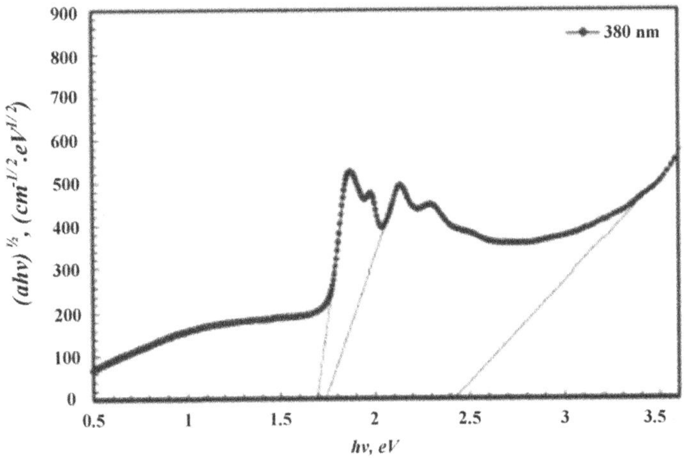

Figure 3.6. *Plot of $(\alpha h\upsilon)^{1/2}$ versus photon energy ($h\upsilon$) for the pentacene thin film of thickness 380 nm*

In the example displayed in Figure 3.6, the values of the corresponding energy gaps for pentacene are 1.68, 1.75 and 2.35 eV, respectively. This characteristic behavior of indirect transitions in organic semiconductor materials corresponds to the transition process in non-crystalline materials [KUM 00]. It should be noted that the first energy value is the optical gap E^{opt}_{g} corresponds to the onset of optical absorption and formation of a bound electron–hole pair, or Frenkel exciton [TSI 02], but the last energy value is the fundamental energy gap (energy gap between valence band "π-band" and conduction band "π*-band" [ZHO 03]), and the values between them may be due to impurities energy levels.

3.2.2. *Photoluminescence*

Photoluminescence describes the process in which a medium is photoexcited by absorption of a high-energy photon, followed by the creation of an exciton. The exciton then relaxes to the lowest ground state. The relaxation of the above exciton can be either radiative, resulting in the emission of a lower energy photon, or non-radiative with no contribution to the PL signal.

Basic photoluminescence and photoluminescence quantum yield measurement performed on emissive material allow us to estimate OLED color, emission spectra and efficiency to generate photon per created exciton. These experiments could be performed on a thin-film sample or in solution. But, it is well-known that the position of the absorption bands of thin film emissive material and emissive material in solution varies with the substrate and with the solvent [BAL 51], respectively.

3.2.2.1. *Basic photoluminescence*

Basic photoluminescence only requires an excitation pump and a spectrometer. Such an excitation pump can be generated by a light source, e.g. lamp, a laser, a LED, or produced by a thin luminescent film, after absorption of an excitation radiation.

Photoluminescence measurement provides an approximate OLED emitted spectrum weighted by the OLED optical out coupling.

The OLED optical out coupling could be estimated from photoluminescence spectra in tuning stack thicknesses, if excitons formation and relaxation are confined within a well-defined region. To realize this confinement, for example, the OLED stack could include carrier blocking layers [FER 05]. Angular photoluminescence measurement can be performed to more precisely quantify the optical out coupling. (the emission from organic LEDs cannot be assumed to be Lambertian, i.e. the flux per solid angle varies as the cosine of the angle from the normal).

3.2.2.2. *Photoluminescence quantum yield measurement*

Photoluminescence quantum yield measurement requires collecting all photo emitted flux, independently of light angular distribution and/or on emission wavelength. These two issues are solved by the use of an integrating sphere and calibrated photodetector.

An integrating sphere, as shown in Figure 3.7, is hollow and coated on its internal surface with a diffusely reflecting material. Photons created within the sphere are channeled, after reflections from the surface, to an exit port. The intensity detected at the exit is directly proportional to the total light flux created within the sphere. It does not depend on the angular distribution of the signal or the orientation of the emitting dipoles in the sample.

To performe absolute photoluminescence quantum yield measurement, the photodetector at the exit must be calibrated in the photoluminescence spectral range.

Photoluminescence quantum yield measurement allows us to measure one factor of the external EL quantum efficiency, the efficiency to generate photon per created exciton (η_{PL}).

– Efficiency definition

The external EL quantum efficiency η_{ext} of an OLED is a product of two factors:

– The optical out coupling η_{opt}: ratio of photons emitted perceived by an external observer per emitted photons.

– The internal EL quantum efficiency η_{int}: ratio of emitted photons per electrons injected.

$$\eta_{ext} = \eta_{opt} \cdot \eta_{int}$$

The internal EL quantum efficiency is itself a function of various factors:

– The internal PL quantum efficiency η_{PL}: ration of photon emitted per created exciton.

– The charge balance γ: ratio of electron per hole available in the OLED emissive layer.

– The injection quantum efficiency η_{inj}: ratio of electron per hole injected in the OLED stack from cathode and anode respectively.

$$\eta_{int} = \eta_{PL} \cdot \gamma \cdot \eta_{inj}$$

$$\eta_{ext} = \eta_{opt.} \cdot \eta_{PL} \cdot \gamma \cdot \eta_{i}$$

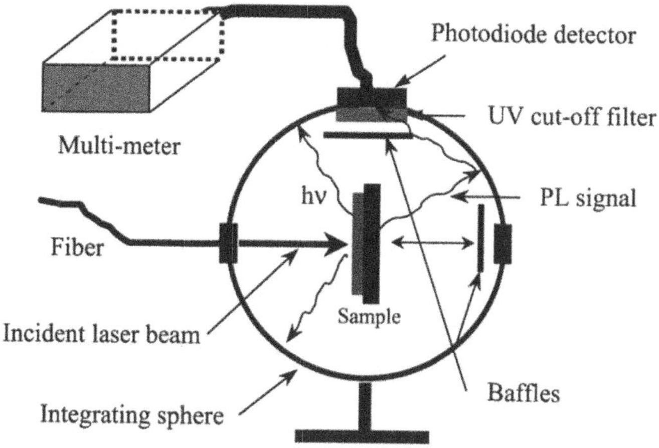

Figure 3.7. *Schematical description of an integrating sphere used to performed PL quantum yield measurements. In the present case, the pump incident beam is guided through the entrance aperture, with an optical fiber, onto the sample placed at the center of the sphere. The reflected signal from the incident beam is directed towards the exit aperture. A baffle placed between the sample and the exit port prevents the PL signal from directly reaching the detector. The PL signal, scattered within the sphere, is measured at the exit with a photodiode, preceded by a cutoff filter*

Since external EL quantum efficiency is a product of various factors, experiments, such as photoluminescence quantum yield measurement, able to estimate one of these factors, are very useful to investigate OLED efficiency.

3.3. Device characterization

3.3.1. *Electrical characterization*

3.3.1.1. *IV characterization (in the dark)*

The relationship between the direct current through an OLED and the voltage across it is called a current–voltage characteristic of the device or IV characterization. This common characterization allows to determine basic parameters of an OLED and to model its behavior in an electrical circuit.

Figure 3.8 shows a typical dark current versus bias voltage (*I–V*) characteristic at room temperature in a semi-logarithmic representation. The device, two OLED emitters, reveals a good diode behavior with rectification ratios between 1e4 and

1e6,and current saturation under reverse bias is observed. From the slope of the semi-logarithmic plot of I versus forward bias the ideality factor can be obtained. However, even in a narrow potential range (2.5–3.5 V), the ideality factors are greater than 3, hence far from ideal as described by different theories.

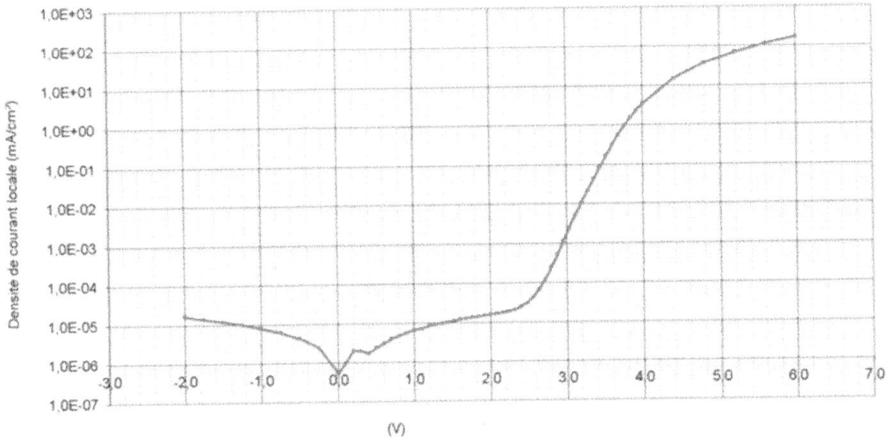

Figure 3.8. *Example of OLED IV characteristics in a semi-logarithmic representation (made in CEA/Leti)*

OLED could generally be modeled by a simple and well-known equivalent circuit displayed in Figure 3.12 resulting in the following equation:

$$I - \frac{V - I.Rs}{Rp} = Is\ exp(\frac{q(V - I.Rs)}{nkT})$$

The equivalent circuit is composed by the Shockley Read Hall diode, a serial resistance and a parallel resistance.

A first method for evaluating serial resistance consists of evaluating, on IV linear representation, the tangent for voltage greater than the diode turn on voltage. A second method consists of evaluating the tangent of the following curve:

$$\frac{\mathrm{I.dV}}{\mathrm{dI}} = \mathrm{f(I)} \qquad \text{with} \qquad \mathrm{f(I)} = \mathrm{Rs.I} + nkT/\mathrm{q}$$

Figures 3.10 to 3.13 display how the ideal diode IV characteristic shown in Figure 3.9 is tuned, in linear and semi-logarithmic representation, by adding a serial resistance, a shunt resistance, serial and shunt resistances and under illumination.

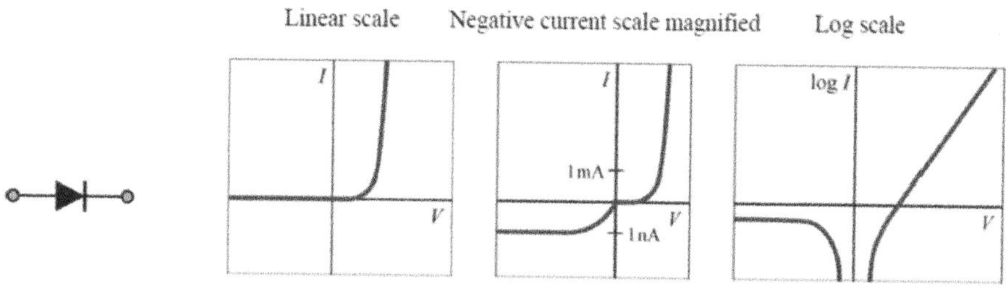

Figure 3.9. *I-V characteristic of an ideal diode (given by the Schockley equation) [SCH 06]*

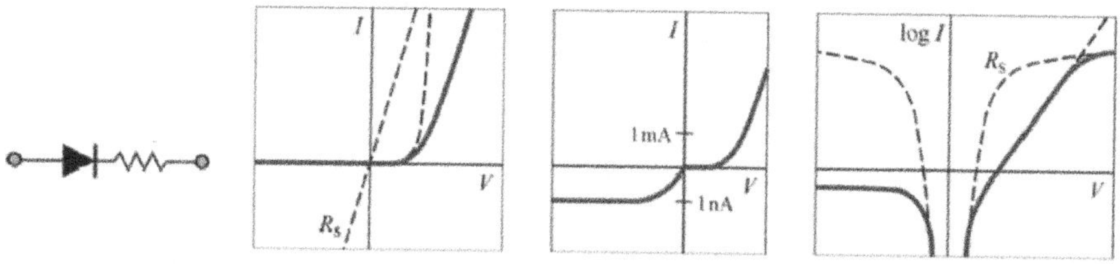

Figure 3.10. *I-V characteristic of a diode with series resistance [BAN 10]*

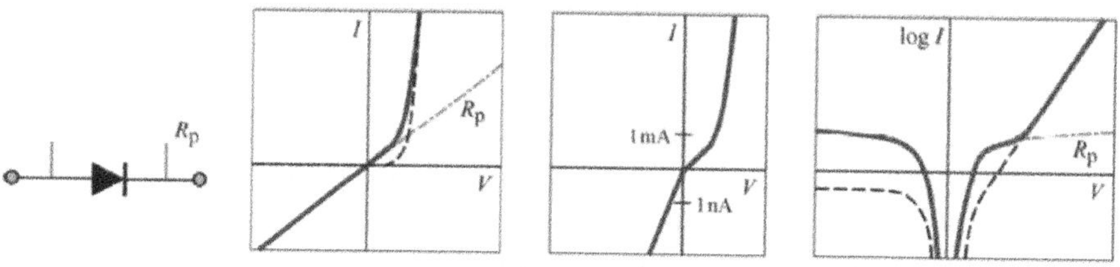

Figure 3.11. *I-V characteristic of a diode with parallel resistance (shunt) [BAN 10]*

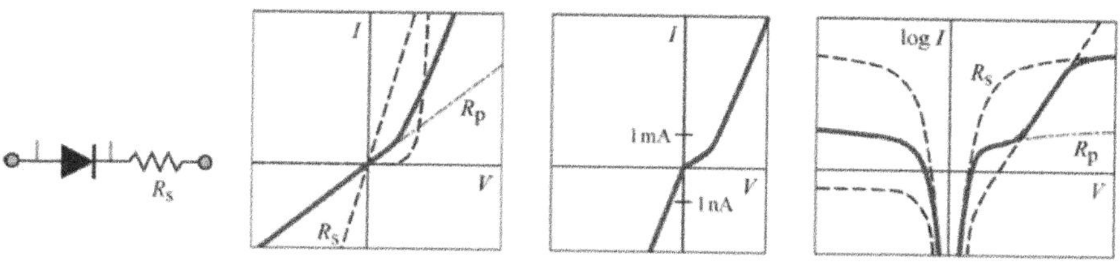

Figure 3.12. *I-V characteristic of a diode with series and parallel resistance (shunt) [BAN 10]*

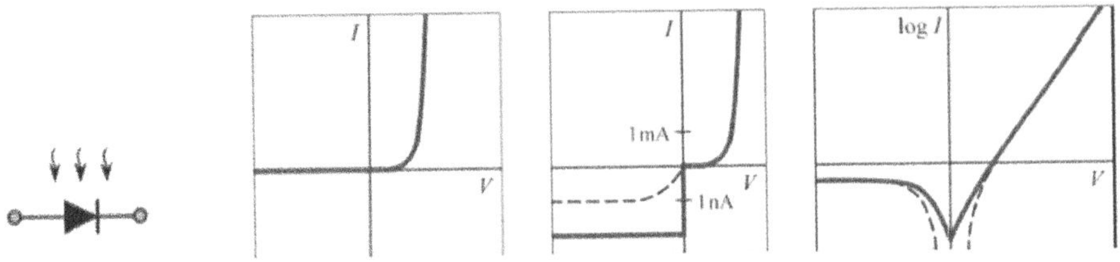

Figure 3.13. *I–V characteristic of a diode with under illumination (with photocurrent) [BAN 10]*

These computed behaviors help the LED user to investigate LED limitations of parasitic effects.

3.3.1.2. *IV characterization under illumination*

Under white-light illumination a photovoltaic effect could be observed (Figure 3.13). The short-circuit current and the open-circuit voltage are given by the interceptions with the axes. Even taking reflection losses into account, the power conversion efficiency of this "solar cell" is generally extremely low. The low short-circuit current points to a significant influence of a serial resistance.

3.3.1.3. *CV and impedance spectroscopy*

Detailed information about the OLED bulk resistance and the OLED junction can be obtained by complex impedance measurements [MAC 87]. For example, Schottky barrier diodes are commonly analyzed by using an equivalent circuit as shown in the insert of Figure 3.14. *Cj* and *Rj* characterize the junction, whereas *Cb* and *Rb* represent the undepleted bulk. The complex plane representation of this equivalent circuit yields two perfect semicircles with diameters corresponding to the ohmic resistances. In the OLED case all quantities are generally field, frequency and temperature dependent and therefore slight deviations from this ideal behavior are expected.

Figure 3.14 shows an example of real and imaginary parts of the complex impedance for different applied bias voltages of the Al/PPV/ITO device.

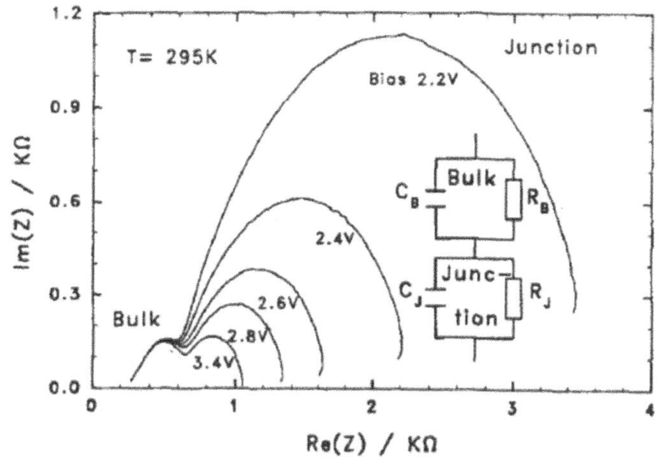

Figure 3.14. *Example of impedance plots under forward bias for an Al/PPV/ITO Schottky diode with an active area of 4 mm^2. The insert shows an equivalent circuit for the Schottky barrier*

The right semicircle falls off rapidly as the device is brought further into forward bias; it represents the junction. The resistive layer of the undepleted bulk is practically not affected and is still present under strong forward bias. At low frequencies and high temperatures the bulk can be neglected and the equivalent circuit reduces to the junction part. Then under reverse bias the measured capacitance corresponds to the junction capacitance, which can be expressed according to the Schottky model as

$$\frac{C}{A} = \left[\frac{q\varepsilon_0\varepsilon_r N_a}{2}\frac{1}{V + V_d}\right]^{1/2}$$

giving a linear variation of $1/C^2$ with the bias voltage in reverse direction. In the previous equation, a represents the device area, V_d is the built-in potential or diffusion voltage and N_a is the ionized acceptor dopant concentration. In Figure 3.15, an example of a plot of $1/C^2$ versus bias voltage measured at a frequency of 100 Hz is shown. The interception of the dashed line with the abscissa gives the diffusion voltage and the slope determines the acceptor dopant concentration. Inserting for the permittivity ε_r a measured value of 4.3, we find for this device a value of $N_a \sim 10^{17}$ cm^{-3} and $V_d \sim 1$ V.

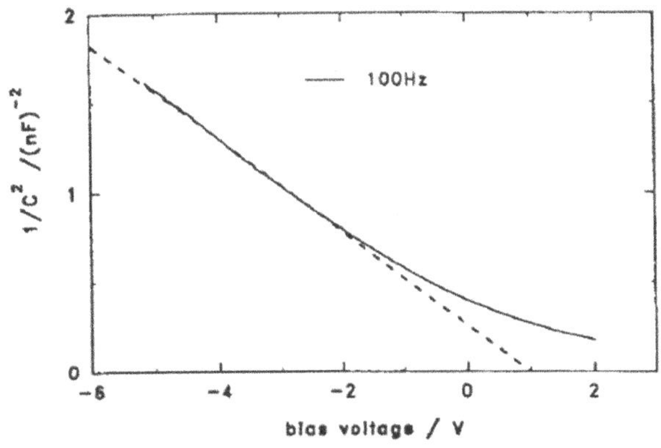

Figure 3.15. *$1/C^2$ versus bias voltage of an Al/PPV/ITO device (area 4 mm^2)*

Thus V_d corresponds to the maximum open-circuit voltage of the solar cells. For an abrupt junction, the Schottky barrier model yields for the depletion width

$$\omega = \left[\frac{2\varepsilon_0\varepsilon_r(V + V_d)}{qN_a}\right]^{1/2}$$

Inserting the values for N_a and V_d allows us to estimate a depletion width of at zero bias.

3.3.2. *Radiometry versus photometry and colorimetry*

3.3.2.1. *Optical flux units: radiometry versus photometry*

Radiometry is the measurement of optical radiation, which include ultraviolet, visible and infrared. Two out of many typical units encountered are watts/m^2 and photons/sec-steradian.

Photometry is the measurement of light, which is defined as electromagnetic radiation detectable by the human eye (wavelength range from about 360 to 830 nm). So, photometry is radiometry weighted by the spectral response of the eye: photopic vision is the vision of the eye under well-lit conditions and scotopic vision under low-light conditions.

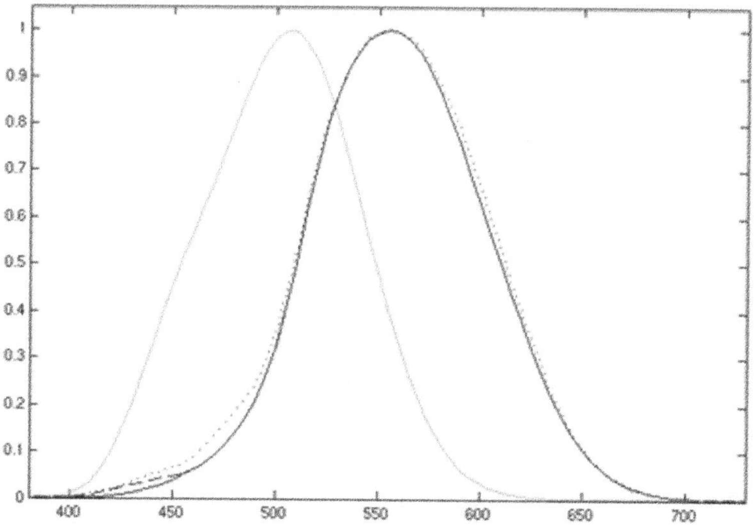

Figure 3.16. *Photopic (black) and scotopic (light gray) luminosity functions. The photopic includes the CIE 1931 standard (solid) discussed in the next section, the Judd–Vos 1978 modified data (dashed), and the Sharpe, Stockman, Jagla and Jägle 2005 data (dotted). The horizontal axis is wavelength in nm*

Typical photometric units include lumens, lux, candelas, etc.

OLED brightness (given in cd m^{-2}) is often used by several authors to estimate the efficiency of their device at a given voltage or current density. For comparison, the brightness of a conventional laptop display reaches approximately 100 cd m^{-2}.

OLED efficiency is commonly expressed in cd/A or in Lum/Watt.

3.3.2.2. *Colorimetry and color coordinates*

The human eye has three types of cells, which sense light, with spectral sensitivity peaks in, respectively, short (420–440 nm, black curve in Figure 3.17), middle (530–540 nm, light gray curve in Figure 3.17), and long (560–580 nm, dark gray curve in Figure 3.17).

Consequently, weighting a total light power spectrum by the individual spectral sensitivities of the three types of cells gives three effective stimulus values; these three values make up a tristimulus specification of the objective color of the light spectrum.

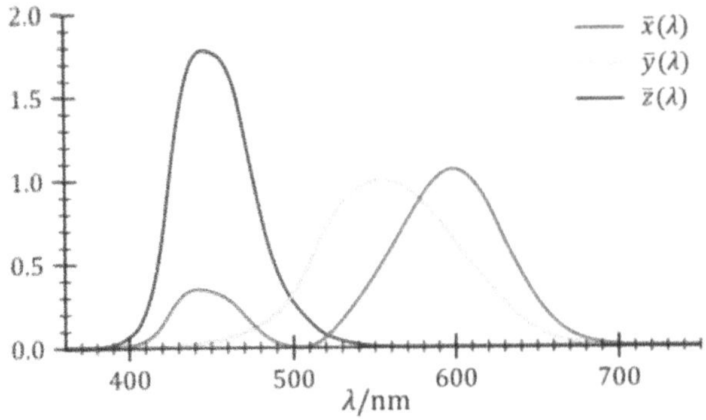

Figure 3.17. *The CIE standard observer color matching functions*

The tristimulus values for a color with a spectral power distribution $I(\lambda)$ are given in terms of the standard observer by:

$$X = \int_{380}^{780} I(\lambda)x(\lambda)d\lambda$$

$$X = \int_{380}^{780} I(\lambda)y(\lambda)d\lambda$$

$$X = \int_{380}^{780} I(\lambda)z(\lambda)d\lambda$$

where λ is the wavelength of the equivalent monochromatic light (measured in nanometers).

The CIE XYZ color space was deliberately designed so that the Y parameter was a measure of the brightness or luminance of a color. The chromaticity of a color was then specified by the two derived parameters x and y, two of the three normalized values, which are functions of all three tristimulus values X, Y and Z:

$$x = \frac{X}{X+Y+Z}$$

$$y = \frac{Y}{X+Y+Z}$$

$$z = \frac{Z}{X+Y+Z} = 1 - x - y$$

The derived color space specified by x, y and Y is known as the CIE xyY color space and is widely used to specify colors in particular in a chromaticity diagram as shown in Figure 3.18.

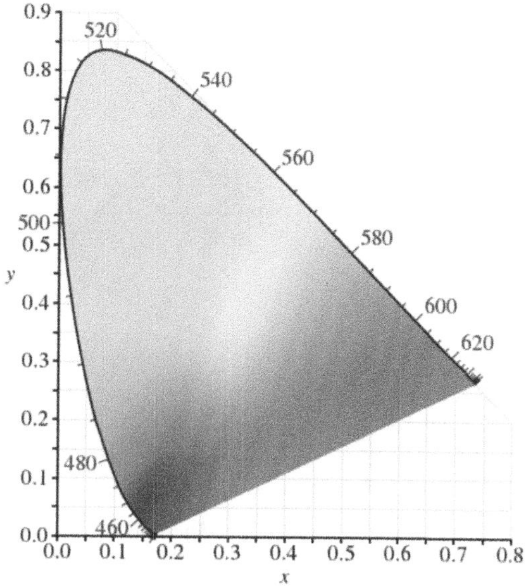

Figure 3.18. *CIE 1931 chromaticity diagram*

Others color spaces are detailed on the International Commission on Illumination (CIE) website (http://www.cie.co.at/) such as CIE 1976 UCS, CIELAB and CIELUV.

3.3.2.3. *Color temperature*

The color temperature of a light source is the temperature of an ideal black body radiator that radiates light of comparable hue to that of the light source. Color temperatures over 5,000 K are called *cool colors* (bluish white), while lower color temperatures (2,700–3,000 K) are called *warm colors* (yellowish white through red). This relation, however, is a psychological one in contrast to the physical relation implied by Wien's displacement law, according to which the spectral peak is shifted toward shorter wavelengths (resulting in a more bluish white) for higher temperatures.

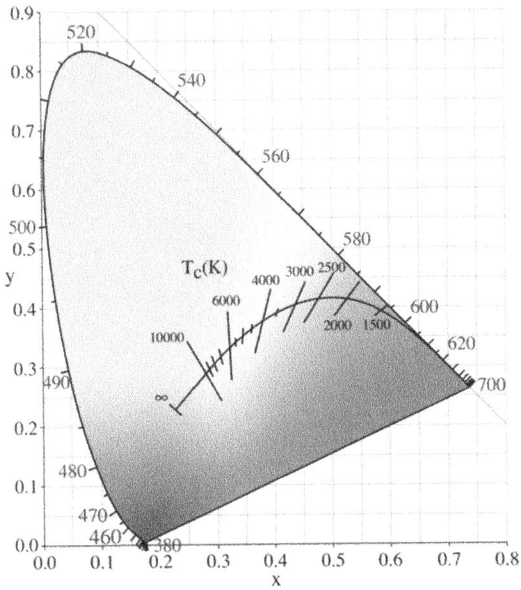

Figure 3.19. *Color diagram on which the black line illustrated the temperature color*

3.3.3. *Electro-optical characterization*

3.3.3.1. *LV characterization*

The relationship between the forward voltage and the luminance emitted by the OLED is called a luminance–voltage characteristic of the device or LV characterization.

Figure 3.20 shows a typical (L–V) characteristic at room temperature in a semi-logarithmic representation. Since emission of one photon requires an electron and a hole, the luminance is proportional to the minority carrier current through the OLED. Furthermore, the assymptotic value of L–V curve, which intercepts the abscissa, is the energy requires by a minority carrier to flow from its injecting electrode to the emissive layer.

As IV characteristic, LV characteristic can be performed at various temperatures. In this way, the hole and electron current are tuned since carrier mobility is a temperature function.

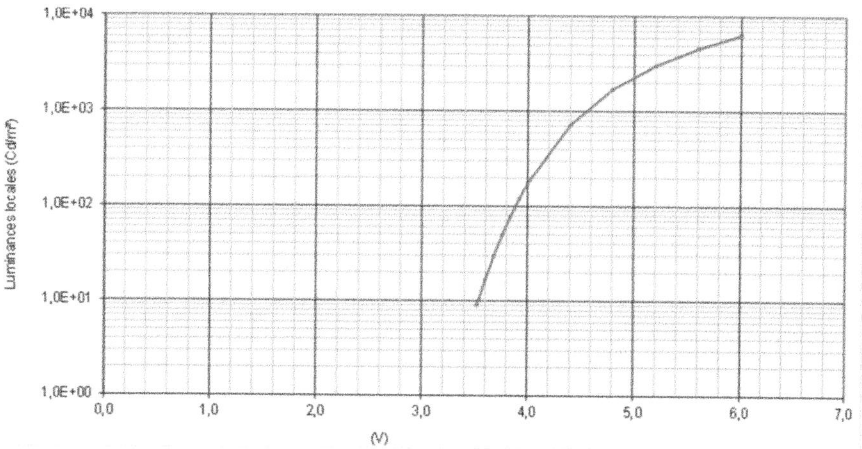

Figure 3.20. *Example of OLED LV characteristics in a semi-logarithmic representation (made in CEA/Leti)*

3.3.3.2. *Electroluminescence spectra*

Electroluminescence differs from photoluminescence in that there is an electrical excitation. Consequently, photoluminescence and electroluminescence spectra can differ due to pump excitation and sample optical out coupling. However, photoluminescence comments and remarks remain relevant for electroluminescence.

In the present example of a two emitters OLED (Figure 3.21), the electroluminescence spectra allows us to distinguish the minority carrier current through the first and second emitter layers. Furthermore, the spectrum shift with voltage translates the carrier distribution shift in the OLED stack.

As photoluminescence measurements, angular electroluminescence measurements can be performed to evaluate the OLED optical out coupling and the color shift.

3.3.3.3. *Pulsed electroluminescence*

Pulsed EL has been studied theoretically and quite extensively experimentally [NIK 98]. The experiment consists of applying a rectangular voltage pulse and measuring the transient current and electroluminescence response.

Delay time, decay and rise for current and electroluminescence are determined by the carrier transport through the OLED stack.

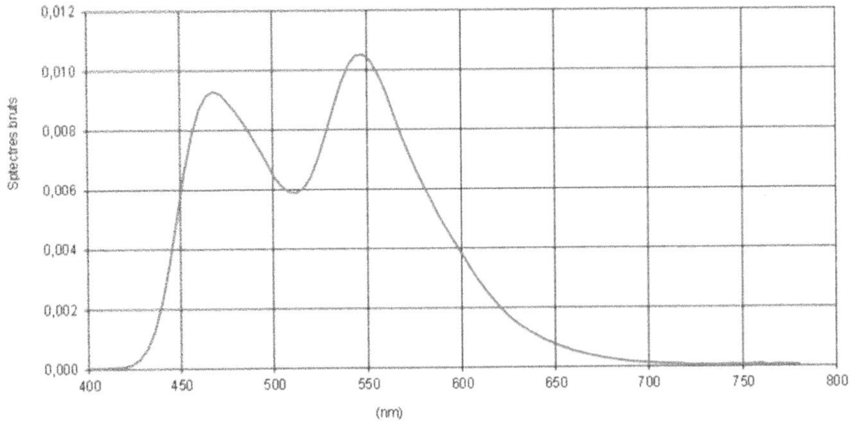

Figure 3.21. *Example of emission spectra of a two-emitter OLED for a voltage (or current) given value (made in CEA/Leti)*

In multilayer OLEDs, charge carrier mobilities can vary by several orders of magnitude from layer to layer. Therefore, as shown in Figure 3.22, the faster carriers penetrate the device and modify the electric field distribution well before the opposite carrier has built up a significant density within the bulk. As a result, a transient current peaks on a very early timescale and is followed by a steep decay prior to equilibrating. At the equilibria, slower carriers reach the emissive layer and can generate excitons and consequently photons. Finally, pulse EL allows us to estimate charge balance γ, one of efficiency factors.

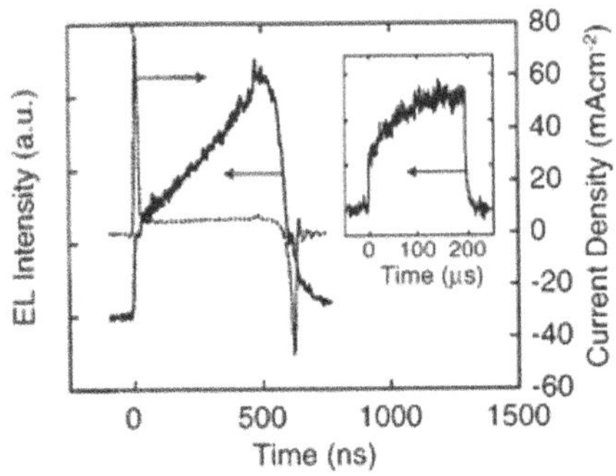

Figure 3.22. *Example of EL and current-density response for 600 ns pulses. The inset shows the EL response for 200 ms pulses*

3.3.3.4. *Efficiency*

Efficiency refers to various ratios:

– current efficiency: ratio of emitted luminance (Cd/m^2) with the current density (A/cm^2);

– power efficiency: ratio of optical emitted power (watt or lumens) with electrical consumed power (watt);

– quantum efficiency: ratio of emitted photons number with injected carriers number, and is expressed in %.

Figure 3.23 shows a comparison between efficiency expressed in Cd/A and Lm/W for fluorescent and phosphorescent emitter OLED.

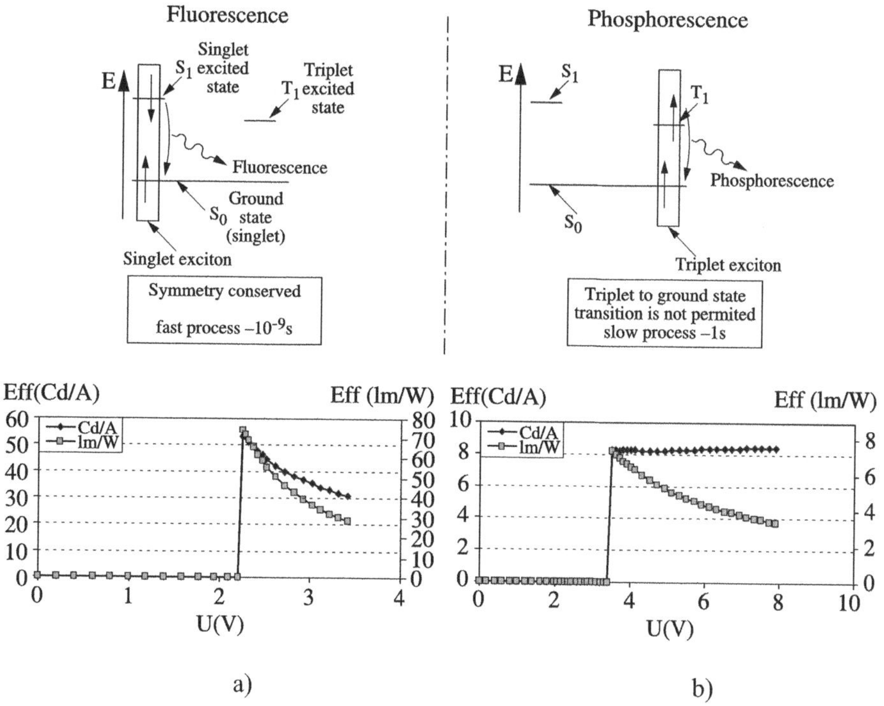

Figure 3.23. *Current efficiency (Cd/A) and power efficiency (Lum/W) of phosphorescent emitter OLED and b) fluorescent emitter OLED*

Fluorescence is a radiative relaxation process from singlet state. Phosphorescence is a radiative relaxation process from triplet states. Fluorescence is an extremely fast process comparative to phosphorescence (singlet exciton state lifetime << triplet exciton state lifetime). However, in fluorescent materials, only decays from singlet state are radiative, while decays from singlet and triplets state are radiative in phosphorescent materials. Consequently, for a similar emission spectra, phosphorescent OLED are significantly more efficient than fluorescent OLED since there are one singlet state and three triplets states per exciton state.

However, as depicted in Figure 3.23, the fluorescent OLED current efficiency remains constant in increasing voltage while phosphorescent OLED current efficiency rapidly decreases. At high injection, phosphorescent emitter OLED suffers from triplet – triplet annihilation process which explains this drop in efficiency.

In conclusion, at low (high) injection or weak (high) luminance, phosphorescent (fluorescent) emitter OLED are more (less) efficient than fluorescent (phosphorescent) emitter OLED. So, the choice between fluorescent and phosphorescent emitter OLED is imposed by the OLED application and the specified luminance.

3.3.4. *Ageing*

Unfortunately, OLED devices suffer from degradation, especially without encapsulation, which explains the argued skepticism about molecular materials: lack of stability and strength required in technological applications.

OLED degradation has been attributed to various mechanisms, including crystallization of organic solids [DO 96a], electrochemical reactions at the electrode/organic interface [DO 96b], migration of ionic species [GAU 96], and electrochemical reactions [AZI 99]. Cathode oxidation and cathode delamination were shown to be largely responsible for the growth of non-emissive spots on the emitting device area [DO 94]. When a light-emitting diode is operated under ambient atmosphere without encapsulation, dark spot formation leads to complete device degradation within a few hours.

Consequently, OLED ageing is required to investigate stability and strength at ambient atmosphere or at accelerated conditions.

3.3.4.1. *Lifetime measurement*

All material and device characterization methods detailed above can be performed to investigate device lifetime. For example, Brutting and coworkers [NOW 09] report that capacitance-frequency analysis is a sensitive tool to monitor charge injection versus time and yields information especially in voltage regimes ($V < V$bi) where current density-voltage measurements are not sensitive.

In this section, we focus on the most common method to estimate lifetime.

Lifetime is commonly defined as the time required at given stress conditions (applied current or voltage) to luminance decays a percentage of initial luminance. For example, lifetime (50%) is time until the luminance decayed to half of the initial luminance.

In the following figures, ageing experiments consist of supplying constant driving current.

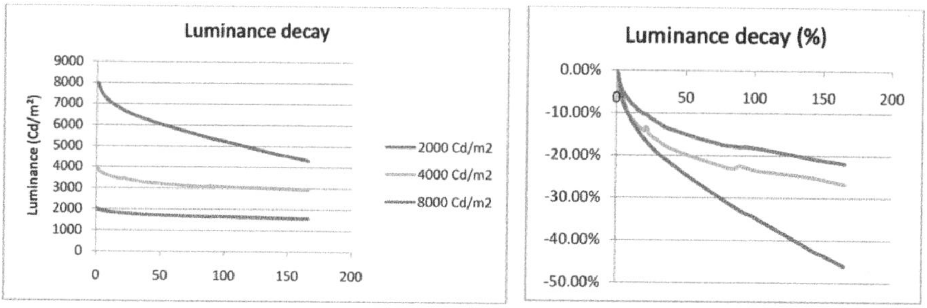

Figure 3.24. *Luminance decay versus time for various initial conditions in a) Cd/m² and b) percentage*

As shown in Figure 3.24, ageing experiments allow us to measure lifetime and to investigate OLED property shift versus time, such as efficiency and voltage, respectively, displayed in Figures 3.25 and 3.26.

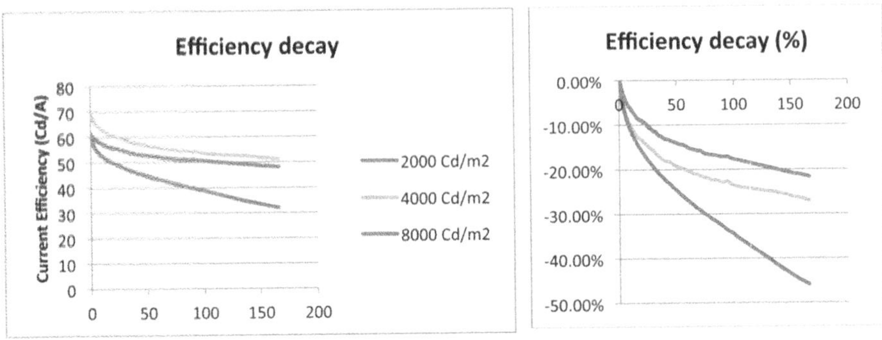

a) b)

Figure 3.25. *Efficiency decay versus time for various initial conditions in a_ Cd/m² and b) percentage*

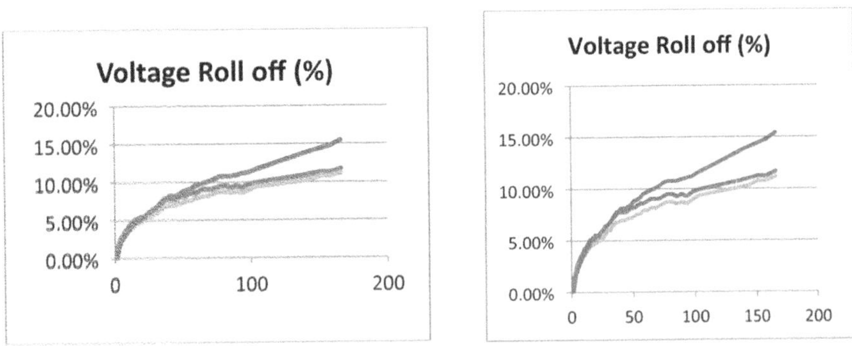

a) b)

Figure 3.26. *Voltage roll off versus time for various initial conditions in a) Cd/m² and b) percentage*

The luminance decay and the voltage roll-off are proportional to the applied driving current. An increase of OLED voltage at constant current means that the OLED serial resistance is increasing. This voltage increase could result in accelerated ageing if the device addressing circuit is a voltage follower circuit.

In the case of constant current drive and in case of appropriate means of encapsulation, OLED lifetime is mainly impacted by the driving current.

3.3.4.1. *Temperature or humidity accelerated ageing*

Lifetime measurement can be performed at a specific atmosphere such as high or low temperature and/or high humidity to investigate OLED sensitivity or robustness in specific atmosphere.

3.4. OLED microdisplay characterization

All the conventional characterizations detailed above can be performed on single μdisplay before or after dicing the backplane (integrated circuit wafer) on the equipment shown below (Figure 3.27). However, since at least one electrode of each OLED of the display is electrically connected to the backplane, OLED properties could be affected by the addressing circuit. The following measures can be made: display luminance, voltage, μdisplay uniformity, μdisplay emission spectra at normal incidence and at several angles, of displays on an integrated circuit wafer (see also section 6.6 for production testing).

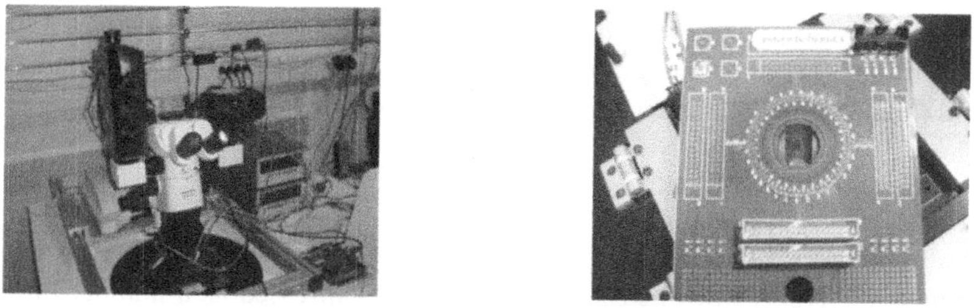

a) b)

Figure 3.27. *a) Testing display prober b) using a probe card allowing us to measure properties, such as microdisplay luminance, voltage, μdisplay uniformity, μdisplay emission spectra spectra at normal incidence and at several angles, of displays on an integrated circuit wafer*

3.4.1. *OLED microdisplay specific measurements*

The applications using μdisplay are mainly near to eye displays such as view finder cameras. Consequently, μdisplay should optimize performances in the context of near-eye viewing.

3.4.1.1. *Resolution*

– Spatial resolution

Display resolution is the human eye's ability to observe the smallest object clearly with distinct boundaries. The smaller the size of the pixel, the higher the resolution and more clear the object in the image. So, the display resolution of a computer monitor or display device is the number of distinct pixels in each dimension that can be displayed. It is usually quoted as width × height, with the units in pixels: for example, "1024 × 768" means the width is 1024 pixels and the height is 768 pixels.

Table 3.1 details the most common display resolutions for computer monitors.

– Spectral resolution

Spectral resolution determines how finely the display can represent differences of color. The spectral resolution is more detailed in contrast, color gamut and uniformity sections.

– Temporal resolution

Temporal resolution determines how the display can resolve events at different points in time. The time resolution used for movies is usually 24 to 48 frames per second (higher than the human eye's temporal resolution).

– Radiometric resolution

Radiometric resolution determines how finely a system can represent or distinguish differences of intensity, and is usually expressed as a number of levels (gray levels) or a number of bits, for example 8 bits or 256 levels that is typical of computer image files. The higher the radiometric resolution, the better subtle differences of intensity can be represented. The radiometric resolution is more detailed in contrast, color gamut and uniformity sections.

3.4.1.2. *Luminance*

A typical computer display emits between 50 and 300 cd/m^2. The μdisplay applications require more or less the same high luminance but with a small size, a high-density pixel, even if they are unaffected by ambient illumination.

Acronym	Aspect ratio	Width	Height
VGA	04:03	640	480
SVGA	04:03	800	600
WSVGA	~17:10	1,024	600
WXGA	16:09	1,280	720
XGA	04:03	1,024	768
WXGA	05:03	1,280	768
HD	~16:9	1,360	768
HD	~16:9	1,366	768
WXGA	16:10	1,280	800
XGA+	04:03	1,152	864
WXGA+	16:10	1,440	900
HD+	16:09	1,600	900
SXGA– (UVGA)	04:03	1,280	960
	03:04	768	1024
SXGA+	04:03	1,400	1050
WSXGA+	16:10	1,680	1050
FHD	16:09	1,920	1080
QWXGA	16:09	2,048	1152
UXGA	04:03	1,600	1200
WUXGA	16:10	1,920	1200
WQHD	16:09	2,560	1440
WQXGA	16:10	2,560	1600

Table 3.1. *Most common display resolutions in the first half of 2012*

So, to achieve high luminance, μdisplays should optimize:

– Angular emission. Instead of 180° viewing angle, the emission directivity must be limit to less than 30° in order to minimize optical losses and consequently maximize μdisplay efficiency.

– Emissive area in order to limit the OLED current density or voltage driving. Consequently, μdisplay efficiency (see Figure 3.23) and lifetime (see Figure 3.24) are maximized, in particular for phosphorescent OLED μdisplay.

3.4.1.3. *Color gamut*

The color gamut is the subset of colors, that can be accurately represented by a display.

Figure 3.28 illustrates an example of a chromaticity diagram with a triangle inside expressing the color gamut of a three primary color display. Each corner of the triangle is defined by the color coordinate of one of the three primary colors of the display. Out of gamut colors (outside the triangle) cannot be expressed by the display.

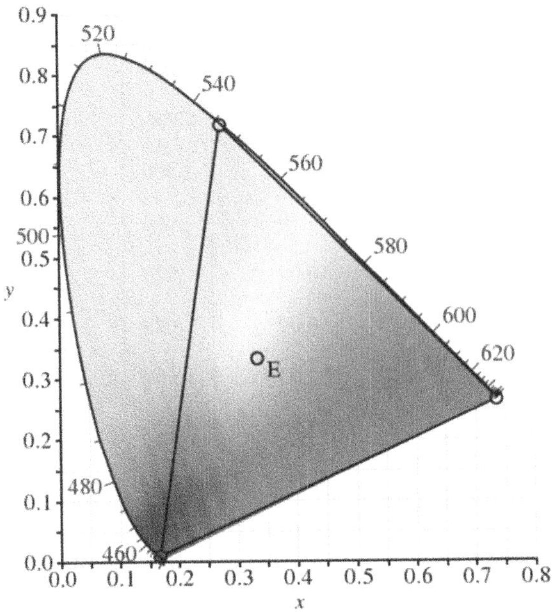

Figure 3.28. *Example of a three primary color display color gamut*

Color gamut standards are provided for example by NTSC, National Television System Committee (http://en.wikipedia.org/wiki/NTSC).

3.4.1.4. *Contrast*

Contrast in visual perception is the difference in appearance of two or more parts of a display seen simultaneously or successively.

The "luminance contrast" is the ratio between the higher luminance and the lower luminance that define the feature to be detected. Since the OLED display is an emissive display, the theoretical contrast could be infinite. However, the luminance contrast is lowered by addressing circuit leakage current and crosstalk. The crosstalk is an undesired emission from a pixel to the adjacent one, there can be an undesired luminance and/or color shift.

Figure 3.29. *Example of pattern useful to investigate display luminance contrast: luminance differences between colored and black parts of a display*

The "color contrast" is the color difference between two parts of a display with equal luminance. Such a color contrast can be described by a distance in a suitable chromaticity system (e.g. CIE 1976 UCS, CIELAB, CIELUV) but not in CIE 1931. In the CIE 1931 color space (see Figure 3.30) the tolerance contours are defined by the MacAdam ellipse, which holds luminance fixed.

MacAdam ellipses (Figure 3.30) refer to the region on a chromaticity diagram, which contains all colors that are indistinguishable to the average human eye, from the color at the center of the ellipse.

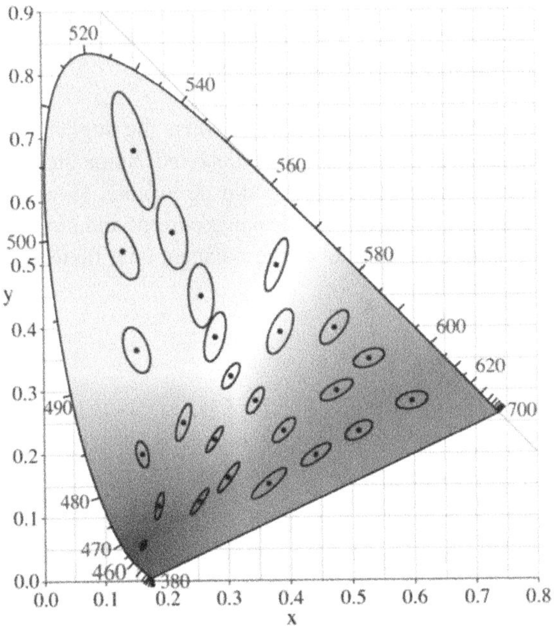

Figure 3.30. *MacAdam diagram in the CIE 1931 color space*

As we can be observed in Figure 3.30, the ellipses denoting the tolerance contours vary in size. It is partly due to this non-uniformity that CIELUV and CIELAB were created.

3.4.1.5. *Uniformity*

Uniformity as contrast is also the difference in appearance (color or luminance) of two or more parts of a display seen simultaneously but which may express the same appearance.

As contrast, uniformity distinguishes “luminance uniformity” and “color uniformity”.

Figure 3.31 shows an example of a result of a uniformity measurement.

Figure 3.31. *Uniformity measurement*

Several factors can explain weak color or luminance uniformity.

The first factor can be the OLED degradation and probably an inefficient encapsulation.

The second factor can be related to the OLED deposition itself, for example, a small thickness variation between two pixels desined to be the same color can modify the optical outcoupling, so the luminance and/or the color. A small electroluminescent layer doping variation, especially for fluorescent emitters, can drastically decrease the OLED efficiency and consequently the luminance at a set driving current.

The third factor can be attributed to the addressing circuit itself.

Adressing circuits are composed of at least two transistors, one of which acts as OLED current source (see Figure 3.32) or voltage supplier.

In the case of Figure 3.32, the OLED driving current is supplied by transistor T2. Since transistor characteristics are not uniform on the backplane, the supplied current can be shifted from one pixel to an other.

3.4.1.6. *Display lifetime*

As detailed in previous sections, different characteristics can be shifted during the μdisplay use, such as:

– an OLED voltage roll-off;

– an addressing circuit voltage roll-off;

– the growth of dark spots;

– an unacceptable color contrast;

– an unacceptable luminance contrast;

– an unacceptable color uniformity;

– an unacceptable luminance uniformity, etc.

Since each of these characteristic shifts could be the µdisplay lifetime limiting factor, no rule can be defined. The μ display buyer or supplier will have to perform their own ageing experiments to check if the device meets their application requirements.

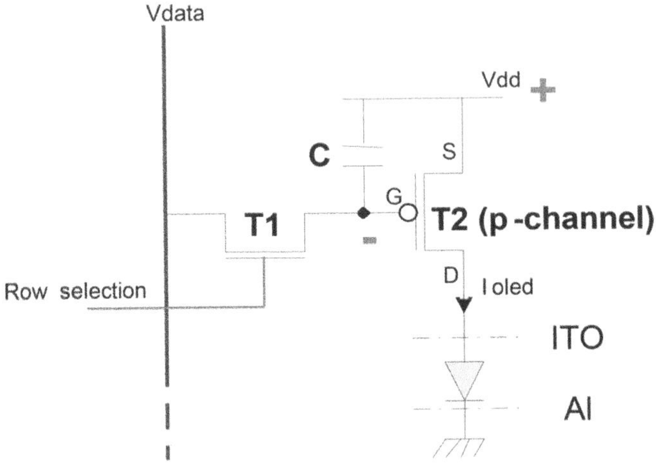

Figure 3.32. *Example of pixel addressing circuit connected to an OLED. Transistor T1 acts as a selector (switch transistor), transistor T2 as a current source (driving transistor)*

3.5. Bibliography

[ABD 06] ABDEL-AZIZ M.M., YAHIA I.S., WAHAB L.A., *et al.*, *Applied Surface Science*, vol. 252, pp. 8163–8170, 2006.

[AZI 98] AZIZ H., POPOVIC Z., XIE S., *et al.*, *Applied Physics Letters*, vol. 72, p. 756, 1998.

[AZI 99] AZIZ H., POPOVIC Z.D., HU N.X., *et al.*, *Science*, vol. 283, p. 1900, 1999.

[BAN 10] BANGE S., SCHUBERT M., NEHER D., *Physical Review B*, vol. 81, p. 035209, 2010.

[BAL 51] BALDER G.M., PEARCE R.S., *Spectrochimica Acta*, Butterworth-Springer, London, vol. 4. pp. 280–283, 1951.

[BOR 93] BORSENBERGER P.M., WEISS D.S., *Organic Photoreceptors for Imaging Systems*, Dekker, New York, 1993.

[CAG 09] CAGLAR M., ILICAN S., CAGLAR Y., *et al.*, *Applied Surface Science*, vol. 255, p. 4491, 2009.

[CAM 00] CAMPBELL A.J., BRADLEY D.D.C., ANTONIADIS H., *et al.*, *Applied Physics Letters*, vol. 76, p. 1734, 2000.

[CAM 01] CAMPBELL A.J., BRADLEY D.D.C., ANTONIADIS H., *Synthetic Metals*, vol. 122, p. 161, 2001.

[CHE 00] CHEN B., LEE C.S., LEE S.T., *et al.*, *Japanese Journal of Applied Physics*, vol. 39, p. 1190, 2000.

[CHE 09] CHEN H.Y., HOU J., ZHANG S., *et al.*, *Nature Photonics*, vol. 3, p. 649, 2009.

[CHI 11] CHIBA T., NAKAYAMA K., PU Y.J., *et al.*, *Chemical Physics Letters*, vol. 502, p. 118, 2011.

[COL 93] COLLINS R.A., KRIE A., ABASS A.K., *Thin Solid Films*, vol. 299, p. 113, 1993.

[CON 94] CONDO P.D., JOHNSTON K.P., "In situ measurement of the glass transition temperature of polymers with compressed fluid diluents", *Journal of Polymer Science Part B: Polymer Physics*, vol. 32, no. 3, pp. 523–533, February 1994.

[DAV 97] DAVIDS P.S., CAMPBELL I.H., SMITH D.L., *Journal of Applied Physics*, vol. 82, p. 6319, 1997.

[DEB 04] DE BOER R.W.I., JOCHEMSEN M., KLAPWIJK T.M., *et al.*, *Journal of Applied Physics*, vol. 95, p. 1196, 2004.

[DO 96a] DO L., HAN E., YAMAMOTO N., *et al.*, *Molecular Crystals and Liquid Crystals*, vol. 280, p. 373, 1996.

[DO 96b] DO L.M., OYAMADA M., KOIKE A., *et al.*, *Thin Solid Films*, vol. 273, p. 209, 1996.

[DO 94] DO L.M., HAN E., NIIDOME Y., *et al.*, *Journal of Applied Physics*, vol. 76, p. 5118, 1994.

[ELK 03] EL-KORASHY A., EL-ZAHED H., RADWAN M., *Physica B*, vol. 334, p. 53, 2003.

[ELN 04] EL-NAHASS M.M., ABD-EL-RAHMAN K.F., FARAG A.A.M., *et al.*, *International Journal of Modern Physics B*, vol. 18, p. 421, 2004.

[ELN 10] EL-NAHASS M.M., YOUSSEF T.E., *Journal of Alloys and Compounds*, vol. 503, p. 86, 2010.

[ELN 11] EL-NAHASS M.M., AMMAR A.H., FARAG A.A.M., *et al.*, *Solid State Science*, vol. 13, p. 596, 2011.

[ESW 11] ESWARD T., KNOX S., JONES H., *et al.*, *Journal of Applied Physics*, vol. 109, p. 093707, 2011.

[FAD 09] FADEL M., FAYEK S.A., ABOU-HELAL M.O., *et al.*, *Journal of Alloys and Compounds.*, vol. 485, pp. 604–609, 2009.

[FER 05] FÉRY C., CINÁ S., DOYEUX H., *et al.*, *Materials Research Society Symposium Proceedings*, vol. 871, pp. 329–33, 2005.

[GAU 96] GAUTIER E., LORIN A., NUNZI J.M., *et al.*, *Applied Physics Letters*, vol. 69, p. 1071, 1996.

[GEN 03] GENEVICIUS K., OSTERBACKA R., JUSKA G., *et al.*, *Synthetic Metals*, vol. 137, p. 1407, 2003.

[GIL 72] GILL W.D., *J. Appl. Phys.*, vol. 43, p. 5033, 1972.

[GRU 14] GRUCELA-ZAJAC M., FILAPEK M., SKORKA L., *et al.*, *Synthetic Metals*, vol. 188, pp. 161–174, February 2014.

[HOR 99] HOROWITZ G., HAJLAOUI R., FICHOU D., *et al.*, *Journal of Applied Physics*, vol. 85, p. 3202, 1999.

[HOR 00] HOROWITZ G., HAJLAOUI M.E., HAJLAOUI R., *Journal of Applied Physics*, vol. 87, p. 4456, 2000.

[HOR 10] HOROWITZ G., *Advances in Polymer Science*, vol. 223, p. 113, 2010.

[ISH 04] ISHII H., HAYASHI N., ITO E., *et al.*, *Physica status solidi (a)*, vol. 201, no. 6, pp. 1075–1094, 2004.

[JUS 00a] JUŠKA G., ARLAUSKAS K., VILIU NAS M., *et al.*, *Physical Review Letters*, vol. 21, p. 4946, 2000.

[JUS 00b] JUŠKA G., ARLAUSKAS K., VILIU NAS M., *et al.*, *Physical Review B*, vol. 62, p. R16235, 2000.

[JUS 00c] JUŠKA G., GENEVICIUS K., VILIU NAS M., *et al.*, *Journal of Non-Crystalline Solids*, vol. 266–269, p. 331, 2000.

[JUS 02] JUŠKA G., GENEVICIUS K., ARLAUSKAS K., *et al.*, *Physical Review B*, vol. 65, p. 233208, 2002.

[KAM 02] KAMATH G.B., JOSEPH C.M., MNON C.S., *Materials Letters*, vol. 57, p. 730, 2002.

[KHA 10] KHAN S.A., AL-HAZMI F.S., AL-HENITI S., *et al.*, *Current Applied Physics*, vol. 10, pp. 145–152, 2010.

[KIM 09] KIM C., FACCHETTI A., MARKS T.J., *Journal of American Chemical Society*, vol. 131, pp. 9122–9132, 2009.

[KUM 00] KUMAR G.A., THOMAS J., GEORGE N., *et al.*, *Physics and Chemistry of Glasses*, vol. 41, no. 89, 2000.

[LAM 70] LAMPERT M.A., MARK P., *Current Injection in Solids*, Academic Press, New York, 1970.

[LEB 97] LEBEDEV E., DITTRICH T., PETROVA-KOCH V., *et al.*, *Applied Physics Letters*, vol. 71, p. 2686, 1997.

[LI 11] LI W., LEE T., OH S.J., *et al.*, *ACS Applied Materials and Interfaces*, vol. 3, p. 3874, 2011.

[LIU 07] LIU C.Y., CHEN S.A., *Macromolecular Rapid Communications*, vol. 28, p. 1743, 2007.

[MAL 99] MALLIARAS G.G., SCOTT J.C., *Journal of Applied Physics*, vol. 85, p. 7246, 1999.

[MAC 87] MACDONALD R., *Impedance Spectroscopy-Emphasizing Solid Materials and Systems*, John Wiley, 1987.

[MAC 01] MACDIARMID A.G., *Rev. Mod. Phys.*, vol. 73, p. 701, 2001.

[MAR 99] MARTENS H.C.F., BROM H.B., BLOM P.W.M., *Physical Review B*, vol. 60, p. R8489, 1999.

[MAT 09] MATSUSHIMA T., MURATA H., *Applied Physics Letters*, vol. 95, p. 203306, 2009.

[MCE 96] MCELVAIN J., ANTONIADIS H., HUESCHEN M.R., *et al.*, *Journal of Applied Physics*, vol. 80, p. 6003, 1996.

[MOT 40] MOTT N.F., GURNEY R.W., *Electronic Processes in Ionic Crystals*, Clarendon Press, New York, 1940.

[MUR 70] MURGATROYD P.N., *Journal of Physics D: Applied Physics*, vol. 3, p. 151, 1970.

[MOZ 05] MOZER A.J., SARICIFTCI N.S., PIVRIKAS A., *et al.*, *Physical Review B*, vol. 71, p. 035214, 2005.

[NAR 95] NARIOKA S., ISHII H., YOSHIMURA D., *Applied Physics Letters,* vol. 67, no. 13, 25 September 1995.

[NIK 98] NIKITENKO V.R., TAK Y.-H., BASSLER H., *Journal of Applied Physics*, vol. 84, p. 2334, 1998.

[NOR 04] NORBERT K., "High purity organic molecular crystals organic crystals", *Germanates, Semiconductors Crystals*, vol. 4, pp. 1–100 1980.

[NOW 09] NOWY S., REN W., WAGNER J., *et al.*, *Proceedings of SPIE*, vol. 7415, pp. 74150G–2, San Diego, CA, 2 August 2009.

[OAN 04] OANA D. Jurchescu, JACOB BAAS, THOMAS T.M. PALSTRA, "Effect of impurities on the mobility of single crystal pentacene", *Applied Physics Letters,* vol. 84, p. 3061, 2004.

[PAL 93] PAL U., SAMANTA D., GHORI S., *et al.*, *Journal of Applied Physics*, vol. 74, no. 10, p. 6368, 1993.

[PER 02] PERCEC V., GLODDE M., BERA T.K., *et al.*, *Nature*, vol. 417, p. 384, 2002.

[PIN 99] PINNER D.J., FRIEND R.H., TESSLER N., *Journal of Applied Physics*, vol. 86, p. 5116, 1999.

[POP 03] POPLAVSKYY D., NELSON J., *Journal of Applied Physics*, vol. 93, p. 341, 2003.

[SAV 99] SAVVATEEV V., YAKIMOV A., DAVIDOV D., *Advanced Materials*, vol. 11, p. 519, 1999.

[SCH 06] SCHUBERT F.E., Light-emitting diodes and solid-sate lighting, 2006.

[SCH 75] SCHER H., MONTROLL E.W., *Physical Review B*, vol. 12, p. 2455, 1975.

[SHI 01] SHIRAKAWA H., *Reviews of Modern Physics*, vol. 73, p. 713, 2001.

[SCO 99] SCOTT J.C., RAMOS S., MALLIARAS G.G., *Journal of Imaging Science and Technology*, vol. 43, p. 234, 1999.

[SPE 69] SPEAR W.E., *Journal of Non-Crystalline Solids*, vol. 1, p. 197, 1969.

[SWA 83] SWAEPOEL R., *J. Phys. E: Sci. Instrum.*, vol. 16, p. 1214, 1983.

[TAN 04] TANASE C., BLOM P.W.M., DE LEEUW D.M., *Physical Review B*, vol. 70, p. 193202, 2004.

[TSI 02] TSIPER E.V., SOOS Z.G., GAO W., KAHN A., *Chem. Phys. Lett.*, vol. 360, p. 47, 2002.

[VER 04] VERES J., OGIER S., LLOYD G., *et al.*, *Chemistry of Materials*, vol. 16, p. 4543, 2004.

[WON 07] WONG W.Y., WANG X.Z., HE Z., *et al.*, *Nature Materials*, vol. 6, p. 521, 2007.

[YAK 07] YAKUPHANOGLU F., BARIM G., EROL I., *Physica B*, vol. 391, p. 136, 2007.

[YAK 07] YAKUPHANOGLU F., KANDAZ M., YARASIR M.N., *et al.*, *Physica B*, vol. 393, p. 235, 2007.

[ZHO 03] ZHOKHAVETS U., GOLDHAHN R., GOBSCH G., *et al.*, *Synthetic Metals*, vol. 138, p. 491, 2003.

4

5-Tools and Methods for Electro-Optic Simulation

4.1. Electro-optic simulation presentation

4.1.1. *Objectives*

As in many other fields of science, the study of organic light-emitting diodes (OLEDs) possesses its own numerical tools to assist the researchers in their efforts to understand and enhance the sophisticated structures. The following assertions will be presented with the objective of describing some practical tools and approaches for microdisplays optimization through modeling and simulation. Thus, not all modeling and/or simulation techniques will be covered in details but quite an in-depth overview of what can be achieved through some R&D tools will be presented.

In order to deal with the study of different physical characteristics of the device, one would have to dispose of electrical and/or optical modeling and simulation, depending on the desired approach. Most of the time, these axes of study are decoupled. Independent optical or electrical solvers can be made or bought from the market, delivering outputs in their respective fields. Each of them solves a set of equations originated from the field they are dealing with, which provides accurate results most of the time. Topics covered in this chapter will be illustrated by examples taken from such solver's outputs.

Since each physical aspect of the device can be dealt with independently up to a certain extent, these solvers represent the simplest way to handle the calculations. Nevertheless, this mono-axial approach of the subject does not give an easy access to hybrid dimensions such as device efficiency or luminance versus voltage.

Chapter written by Karim BOUZID.

Correlation between solver's outputs is then required to proceed further toward electro-optic simulation outputs.

4.1.2. *Potential gains*

Both types of modeling/simulation, and even different models inside each type, are not equivalent in terms of ease of implementation, quantity requirements, cost and complexity, but both still represent a significant investment. Taking a step toward such techniques is a promise for a fast, cost-efficient product development cycle. Countless try outs from DOE can thus be prevented. When established and made reliable, simulation outputs can for instance give access to a set of goal-optimized thicknesses for the OLED stack, hints about changes induced by the variation of one or several material parameters and more.

Accuracy of the results is a recurrent concern when coming to calculated entities. One can roughly say that the accuracy of a simulation is proportional to the investment made on and around the tool. For this reason and in order to be as close to realistic values and as reliable in calculations as possible, techniques presented in this chapter are all based on or related to experiments. Either by fit procedure or simple measurement, all material data definition has an experimental origin.

4.1.3. *Available software solutions*

Among the possible modeling/simulation environments, we can consider the lab-made solutions, often more tunable but less optimized and handy, but also the commercial alternatives. In this chapter, optical output examples will be calculated from the software SETFOS© from FLUXIM and electrical outputs from SimOLED© from SIM4TEC. A third company is also to be mentioned, SILVACO, which proposes an "Organic display" module in its ATLAS© simulation environment. Although these commercial softwares deal with most situations classically encountered, some topics, such as inter-pixel electrical crosstalk, are not covered and require specific lab-made simulators.

4.2. Optical simulation

Optical simulation for OLEDs represents the most common type of simulation. Quite easy to implement, reliable without complex machinery to support it, based on a known and established physical model, this type of simulation is the safest for OLED microdisplays optical optimization through calculation.

Solver operation

The main work of the simulation tool is to optically solve multi-layer stacks. Using the transfer matrix formalism is an effective way to deal with thin multi-layer structures. By computing the reflectivity, the transmittance and the absorptivity, the photonic flux density inside the multi-layer can be calculated. For more details about the transfer matrix formalism, please refer to [YEH 05].

Inputs

In order to launch a simulation, feeding the model with appropriate data is mandatory. Among the inputs asked to be given, here is a summary of the typical ones that can be encountered when dealing with optical simulation: stack layer thicknesses (organic material and metals), n & k index spectrums for each material, emission spectrum shape for each emitting material, their emission intensity, the recombination zone position within the stack and finally the type of dipole orientation that needs to be taken into account.

Getting the inputs

To obtain the numerical values of the model inputs, the following techniques are available. The stack layer thicknesses are controlled during the deposition and were previously calibrated, the n & k index spectrums are measured for each material by ellipsometry, where the exact thickness of the deposited sample has been confirmed by an XRR machine (X-ray reflectivity). The emission spectrum shape has been normalized after being acquired via a spectrophotometer by either electroluminescence (luminescent stack) or photoluminescence (simple thin film deposition or in solution). Emission intensities are set according to the spectrum obtained in the output for a given stack. The output should match the shape of the experimental emission spectrum. The recombination zone position is approximately known by design (which layer) and should be adjusted with the same technique as for the previous parameter, in accordance with the experimental emission spectrum. The type of dipole orientation depends on our material. In our case of evaporated small molecules, the orientation is isotropic. This parameter is important for light-coupling matters and emission characteristics.

4.2.1. *Bottom-emission OLEDs*

Two classical types of OLEDs can be distinguished according to the side from which the light will come out: bottom-emission and top-emission OLEDs (see Figure 4.1). The first kind is generally built onto a transparent substrate, typically glass/ITO or PET, and light will come out through it. The second kind is built on an opaque substrate, could it be a silicon wafer, steel or anything smooth enough in surface for OLED material deposition, light coming out on the opposite side of the substrate.

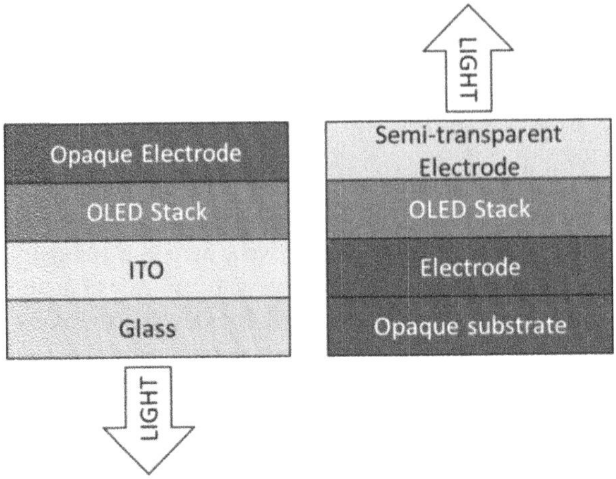

Figure 4.1. *Scheme of a bottom-emission OLED (on the left) and a top-emission OLED (on the right). This, and all other figures in this book, is available in full color at www.iste.co.uk/templier/OLED.zip*

Considering the requirements in terms of inter-pixel spacing and pixel density for microdisplays, bottom emission technology cannot be considered as adequate. Indeed, in bottom emission, transistors addressing the pixels are located in the inter-pixel region. This region's dimensions are reduced almost down to the technologies limits in the case of microdisplays. Also, it is difficult to build an addressing circuit on top of the OLED stack due to temperature limits imposed by OLED materials, low temperature processes only being allowed. For these reasons, bottom emission is not suited for the fabrication of microdisplays. Though, it is interesting to stress that bottom-emission OLEDs are almost free from the microcavity effect, which is due to the high transparency of the anode, allowing a much less constrained design regarding layer's thicknesses for instance.

4.2.2. *Top-emission OLEDs*

4.2.2.1. *Microcavity: existence, impact*

On the other hand, top-emission OLEDs suffer or benefit (depending on the objectives) of a strong microcavity effect [BUL 98]. This phenomenon is due to the electrodes, behaving either as mirrors or semi-mirrors and allowing a certain set of wavelengths to resonate within the device. Wavelengths concerned will see their extraction coefficient promoted, whereas the remaining ones of the spectrum will be attenuated. Microcavities in top-emission OLEDs behave like a low quality-factor Fabry-Perot device and can be characterized with reflectometry. Reflectometry will stress which wavelengths are reflected by the device and which are trapped within. The minimum of the curve indicates at which wavelength the effect is centered. The value of the normalized intensity at this precise wavelength and the width of the reversed peak give information on the strength of the cavity. An experimental example is shown in Figure 4.2 where the cavity is centered on 530 nm. Cavity is tuned via the layer's thickness (modifying the optical distance between electrodes) and the thickness of the electrodes (modifying its reflectance level and thus the quality factor of the microcavity).

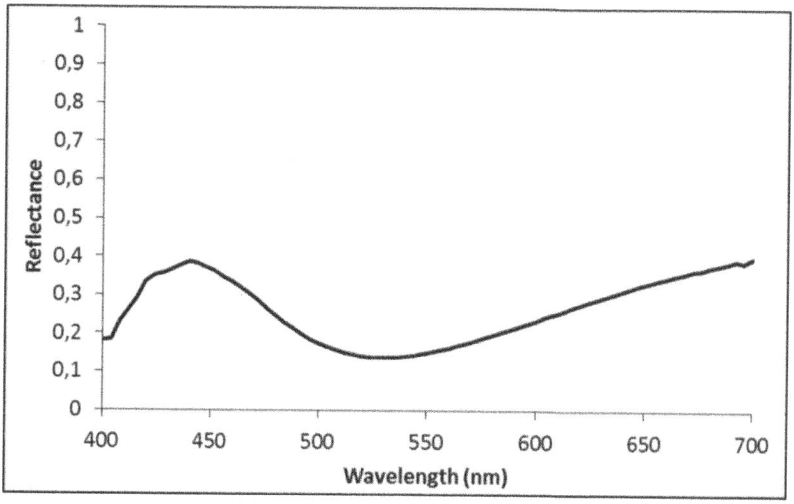

Figure 4.2. *Reflectance characteristic of a 530 nm centered top-emission OLED obtained by reflectometry and a 10 nm Ag semi-transparent electrode*

As stated, microdisplays are fabricated with top emission technologies, thus rendering it very dependent on microcavity effects. When dealing with monochrome displays this can be turned as an advantage, since the effect can be centered on the interesting wavelengths. On the other hand, if we want to make multi-colored displays, an initial white display on which will be added color filters will be needed. The choice of white light emission combined with color filter is imposed by the order of magnitude of the pitch in microdisplays. In this case, the cavity effect complexifies the design of the stack, since several separated wavelengths need to be considered. The following section will illustrate the latter point.

A second impact of working with microcavities is the obligation to place the recombination zone of the OLED roughly in the maximum peaks of the cavity. Resonance does not occur the same way along the cavity, we need to consider the nodes of vibration and the peaks for maximum light extraction. For the spectral domain on which the studies are focused (450–650 nm), the microcavity lengths corresponding to the optical prerequisites are in the range of 150–300 nm, depending on the color aimed at. If we would want to use such dimensions, and work in the fundamental mode of vibration, the recombination zone should be placed in the middle of the cavity, at the maximum vibration peak. Working at the first vibration mode consequently places the recombination zones at one quarter and three quarters positions of the cavity. The fact is that microdisplays need to be electronically addressed with fairly low voltages and with a great responsivity. Using thicknesses as high as 150–300 nm is thus sub-optimal-performance wise, for a microdisplay application. Working at the first mode of vibration is thus preferred, since the vibration properties stay the same with half the stack thickness. In conclusion, for, a microdisplay having half the microcavity length and operating in the first vibration mode and not the fundamental one, the recombination zone is optimally placed in the middle of the stack. Figure 4.3 illustrates the peaks and nodes of vibration in first mode and the optimal positioning of the recombination zone.

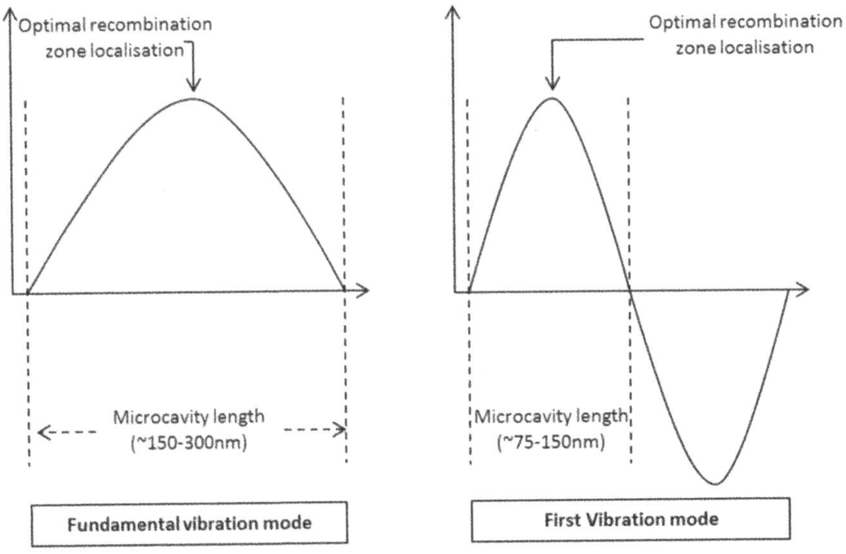

Figure 4.3. *Representation of the fundamental and first vibration mode inside an OLED and their respective optimal recombination zone localization for maximum light extraction*

4.2.2.2. *Dual emitter white-OLED example*

The example treated below explains the tuning of the thicknesses of a blue-yellow dual emitter OLED stack layers in order to obtain a white emission. The electroluminescence spectrums of the two dies are given by the material provider. They are illustrated in Figure 4.4.

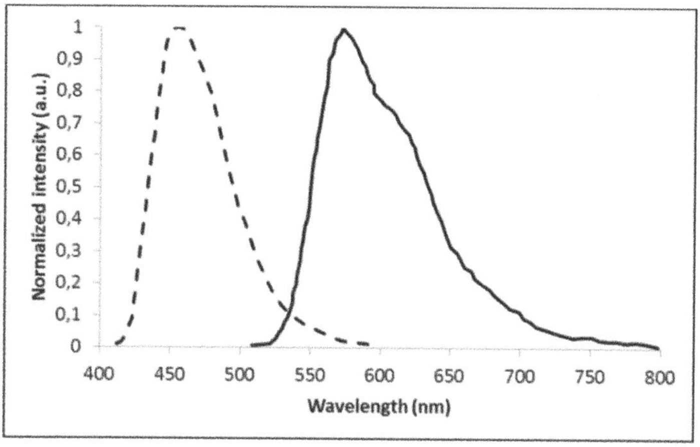

Figure 4.4. *Normalized electroluminescence spectrum of a blue dopant (dashed line) and a yellow dopant (full line)*

For a given electrode, optical length of the cavity and its peak position can be determined from stack total thickness with the help of optical simulation. In order to allow the extraction of both colors from the device, the cavity must be centered in between the characteristic wavelengths of the two colors. Total stack thickness will be designed in order to achieve this goal. Figures 4.5, 4.6 and 4.7 illustrate the emission of the same stack design in which only the total thickness of the stack is varied. Since the OLED is p-i-n type, only p-doped and n-doped regions will be modified, thickness-wise, in order not to affect the electrical properties of the device. The recombination zone will be fairly kept in the middle of the stack. In Figure 4.5, the cavity is centered on the blue emitter. See how the contribution of the yellow emitter is negligible. The same complementary observation can be made when centering the cavity on the yellow emitter. See how in Figure 4.6 the blue component is absent from the emitting spectrum. In Figure 4.7, the cavity is placed at 530 nm, in between the two characteristic wavelengths. Spectrum displayed in Figure 4.7 gives white light. Slightly red or blue shifting the center of the cavity allows tuning the temperature of the white color, by increasing the contribution of yellow or blue. A more yellow component will result in a warmer white and vice versa, more blue will produce a colder blue.

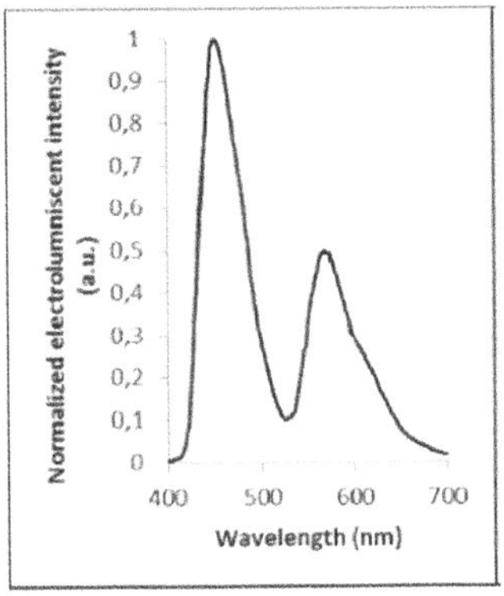

Figure 4.5. *Normalized emission spectrum for blue-centered microcavity*

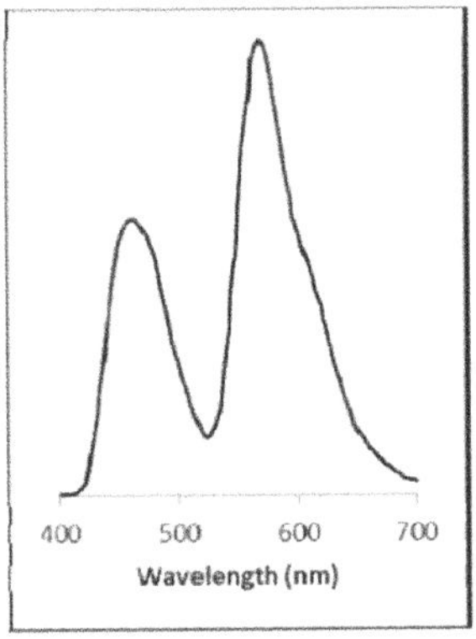

Figure 4.6. *Normalized emission spectrum for yellow-centered microcavity*

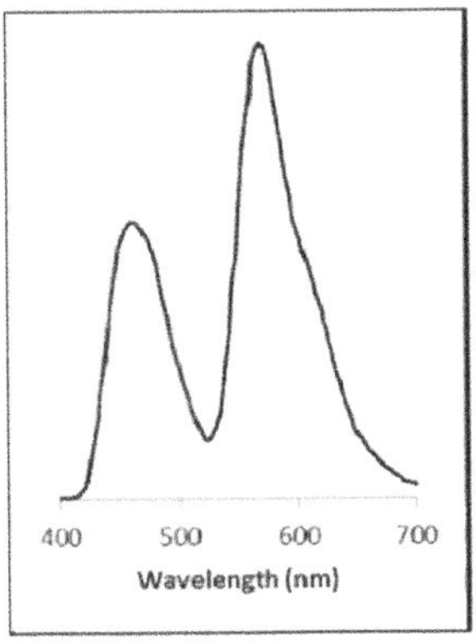

Figure 4.7. *Normalized emission spectrum for 530nm centered microcavity*

4.2.2.3. *Alternative structures for white OLEDs*

Other ways for obtaining white OLEDs for microdisplays exist, but all have drawbacks. The first technique is based on making a blue emitting diode on which is a phosphorescent yellow layer deposited. Due to energy down conversion, white light is emitted in the end, and filters have to be added to obtain red, green or blue (RGB) colors. The main drawback of this technique is to rely on blue emitters that show relatively poor lifetime performances.

The second alternative technique is more difficult to achieve process-wise but presents the advantage of giving maximal luminance for a given stack. The concept relies on having the same white-OLED stack on all the pixels of the microdisplay, but the different RGB pixels would present each a different microcavity length in order to center the cavity on the wavelengths that the pixel is supposed to emit (i.e. RGB). The pixel would then only emit the part of the spectrum that its cavity is centered on and not the whole spectrum. The structure is illustrated in Figure 4.8. To have different microcavities for each pixel, a layer with a different thickness on each pixel has to be inserted between the two electrodes. The material chosen would thus have three important features: it should be transparent not to degrade the emission properties, it should be highly conductive (either hole-or electron-wise) not to degrade the electrical properties and finally it should be patternable in a clean way down to the dimensions of the pixels.

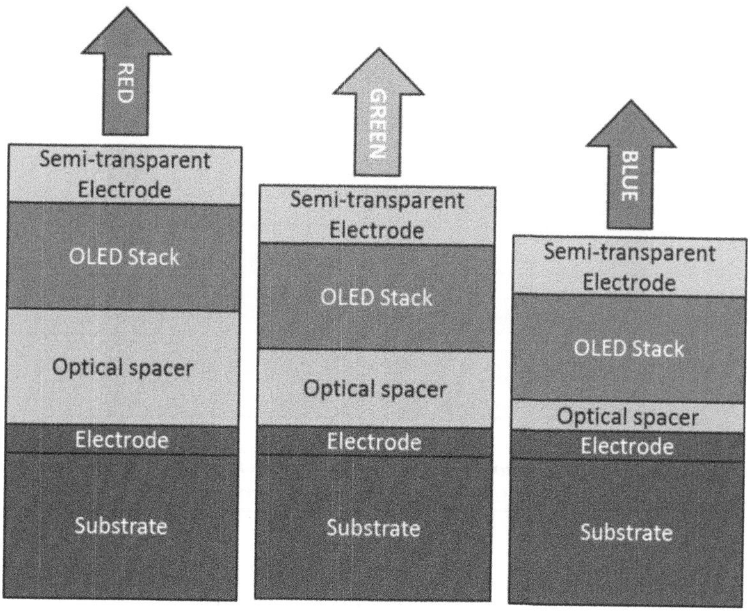

Figure 4.8. *Pixels scheme on how to obtain RGB colors with white OLED and optical spacers*

4.2.2.4. *Radiation pattern and spectral narrowing*

With a typical metallic semi-transparent electrode, the OLED emission is lambertian in the top emission, and no further effort has to be made to obtain wide angle microdisplays. Nevertheless, in some specific cases a narrow monochromatic directive emission is needed, and the example that follows will illustrate how to change the radiation pattern of a yellow OLED.

To restrain the angle of emission and the spectral width, a dielectric mirror can be placed on top of the OLED, right after the cathode. [TOK 99] This dielectric mirror is constituted of low and high refractive index quarter-wave layer pairs. Low refractive index layers will be called L and high refractive index layers will be called H. Figure 4.9 illustrates the two possible structures comprising a dielectric mirror onto a top emission OLED. L-$(HL)_x$ will be called odd structures and $(HL)_x$ will be called even structures.

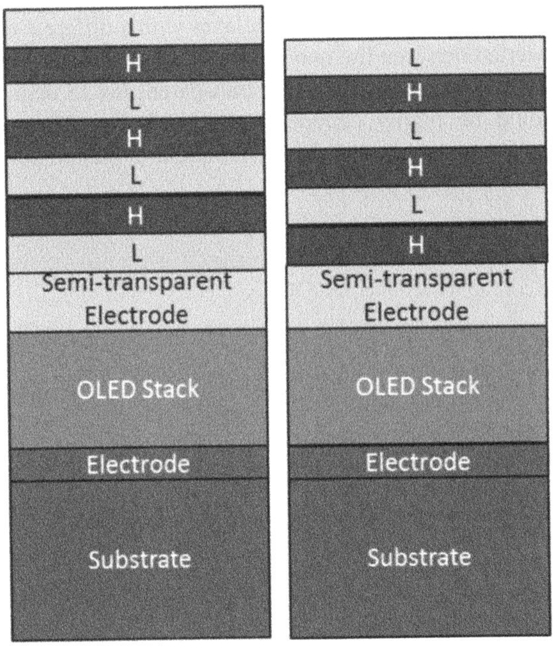

Figure 4.9. *Representation of the two types of dielectric mirror structures on top of an OLED, odd structure (left side) and even structure (right side)*

These two structures are not equivalent in terms of induced radiation pattern, and the difference can be seen in Figure 4.10. For a Lambertian profile, half maximum intensity is reached for an angle of 60°, for an HLHLHL structure, the half maximum intensity angle is 44° and finally the smallest angle is reached for the L-HLHLHL structure with a value of 32°. This value stays nearly the same with lower orders of the dielectric mirror (i.e. L-HLHL and i.e. L-HL).

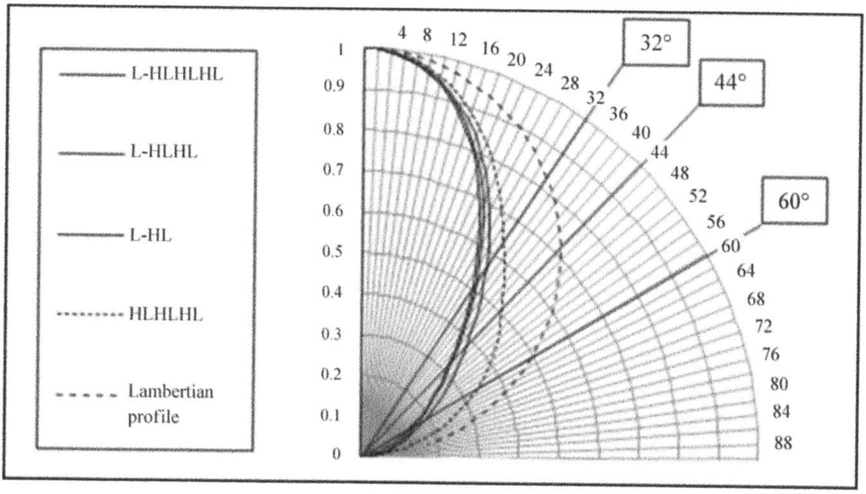

Figure 4.10. *Normalized radiation patterns for even and odd structures and their half maximum angles compared to a Lambertian profile*

The number of L/H pairs also impacts on the emission intensity. In Figure 4.11, three odd structures with different number of pairs are compared to a third order even structure. We can see that the maximum emission intensity is reached for the third order even structure, but at the cost of a wide radiation pattern. Odd structures show less emission intensity than the even structure but a more directive radiation pattern. Concerning the order of the odd structure, the higher the number of pairs is, the lower the intensity will be, while conserving nearly the same radiation pattern. The best solution for this example is thus the one pair odd structure.

To go even further in the spectral narrowing of the emission, the semi-transparent cathode thickness can be increased up to a maximum. A too thin layer would cause not enough selectivity and a too thick layer would cause troublesome absorption, thus an optimum has to be found. Additionally, for each thickness of the cathode, the reflection coefficient of the semi-transparent is changed, and the cavity has to be adapted. This can be adjusted through balanced variation of p and n-doped layer thicknesses. Figure 4.12 displays the spectrums for different cathode thicknesses. We can see an optimum for 25 nm with a FWHM of 30 nm, whereas the original microcavity-free FWHM is 80 nm.

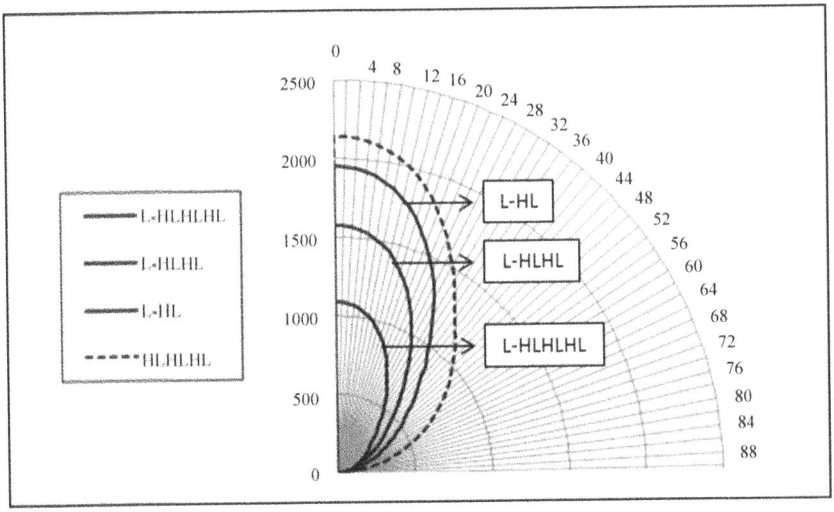

Figure 4.11. *Radiation pattern for different dielectric mirror structures*

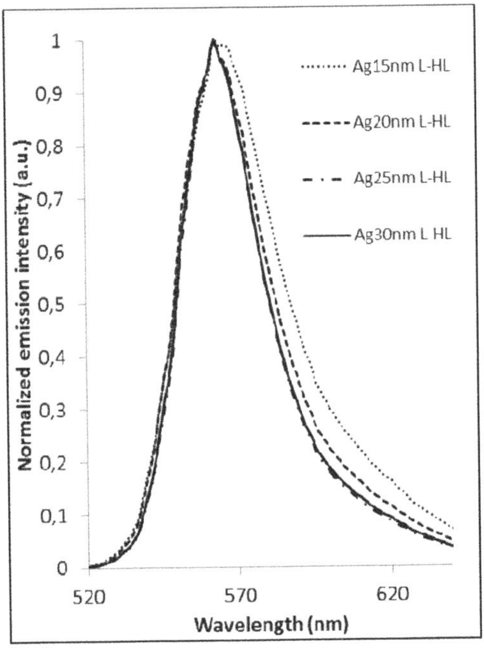

Figure 4.12. *Normalized emission spectrum for different silver electrode thicknesses for a L-HL dielectric mirror structure*

4.3. Electrical simulation

4.3.1. *Potential gain*

Electrical simulations are tougher to set-up than the optical ones. This is mainly due to the difficulty to access material parameters and the accuracy with which we can obtain them. A great effort has to be put into material parameters determination and thus, it is by nature a heavy investment but not at all deprived from interest. With a fully informed material database, we could consider the exploitation of the tool as a sophisticated pre-optimization step of the OLED stack.

4.3.2. *Simulation types*

Two very different kinds of approaches are possible when dealing with electrical simulation. The choice of one over the other is mainly determined by the goal of the study. These two approaches are the Monte-Carlo type simulation and the continuum modeling/simulation and will both be presented, even though a particular focus will be placed on the latter approach.

– Monte-Carlo type simulation

- Operation

The Monte-Carlo method is based on a probabilistic approach to the situation. In the case of OLED simulation, this will intervene, for example, in the consideration of the hopping mechanism between molecules in the computation of the transport properties of the materials.

- Objectives and limitations

This type of approach is interesting when dealing with the simulation of charge transport through the hopping mechanism description and will deliver in-depth, molecularly localized outputs. Computation also happens to be possible in a three-dimensional environment, leading to a spatialized representation of the charge transport within a layer. One of the drawbacks of this method is its long calculation time and its complexity in terms of inputs. Often, Monte-Carlo simulations are not run on personal computers but on clusters, and despite thus, the average computation time for a single simulation run is several days. This is why this approach is better suited for academic purposes and not for device optimization and multi-layer stack electrical behavior prediction.

– Continuum

- Operation

The continuum approach is a method for which equivalents to physical entities or experimental behaviors have been determined. These equivalents, should it be entities, such as field-dependent mobility or trap depth, or whole electrical behavior equations, such as Poisson's equation, are spatially averaged equivalents, for each material. With this in mind, solving the problem in one dimension (1D) should seem legitimate as a simplification procedure. This is especially needed, since the computation time is non-linearly proportional to the number of dimensions the problem needs to be solved in. The way this type of approach should be considered is as a light and fast computing 1D simulation with functional bricks as inputs.

- Inputs

The typical inputs we would need to feed the model with are: the energies of the transport levels of the organic semiconductors (HOMO (highest occupied molecular orbital) and LUMO (lowest unoccupied molecular orbital)), material permittivity, layer thicknesses, built-in voltage, electron and hole mobility (see below for more details on different models of mobility), trap depth and density, electron and hole-wise and finally exciton-type parameters such as singlet/triplet lifetime, quenching cross-section, hopping distance, efficiency of radiative decay, photon/exciton energy, host-guest distance and Foerster radius.

- Mobility models

For continuum modeling, there are at the moment four types of models to consider the mobility of hole and electrons within organic semiconductors. These four models account for different degrees of complexity and dependencies.

1) Poole-Frenkel mobility model

The Poole–Frenkel model is the oldest, simplest and most known empirical model to represent a simple dependence of the mobility versus electric field. It originates from the description of electrical transport in an insulator through localized states. Describing room temperature behavior with this model is fairly easy, since the inputs (zero field mobility and field dependence factor) do not have too strong an overlap in the way it impacts on the transport properties. Nevertheless, this simplicity has a drawback in terms of accuracy when describing low regimes in I–V characteristics or coherence when looking for a large/medium range thickness description. Temperature dependence is not a feature in this model.

Poole–Frenkel definition of carrier mobility: $\mu = \mu_0 \exp(C\sqrt{E})$, where μ_0 is the zero field mobility, C the field dependence factor and E the electric field.

2) Poole-Frenkel-Temperature mobility model

The Poole–Frenkel Temperature model is an evolution of the Poole–Frenkel model, where a temperature dependence factor has been added in the definition of the carrier mobility. The electric field dependence is conserved as for the limitations stated previously.

Poole–Frenkel Temperature definition of carrier mobility: $\mu = \mu_0 \exp\left(-\frac{E_a}{kT} + C\left(\frac{1}{kT} + \frac{1}{kT_0}\right)\sqrt{E}\right)$, where μ_0 is the zero field mobility, C the field dependence factor, E the electric field, T_0 a characteristic temperature and E_a the activation energy.

3) Extended Gaussian Disordered Model (EGDM)

The Extended Gaussian Disordered Model (EGDM) is a model based on Monte-Carlo simulations. The main difference with the previous approaches is the consideration of the disordered nature of organic semiconductors, using distributed transport states. Thus, in addition to electric field and temperature dependence, the charge carrier concentration dependence is taken into account. Further details about the equations can be found in [PAS 05].

4) Extended Correlated Disordered Model (ECDM)

The Extended Correlated Disordered Model (ECDM) is a different version of the EGDM, where the energy states are considered to be correlated. This has for a consequence a higher dependence on the electric field than with the EGDM. Further details about the equations can be found in [VAN 10]. The mobility is also still a variable of temperature and charge of carrier concentration.

- Fit method for data extraction: example of Alq3

In order to perform a simulation, we need material data to feed in the equation solver; and with accurate values if possible, it can only enhance the output accuracy. Unlike optical simulations for which these data values can either be directly measured or intuited, electrical modeling requires particular ways to extract material parameters. For instance, trying to insert in a Poole–Frenkel model a zero-field mobility obtained by a time-of-flight experiment on a thick device (>1 µm) would result in nothing more than a vague approximation concerning the transport properties of this material. In the following section, a technique for material parameter extraction through current voltage characteristic fitting will be presented.

Assuming the goal is to simulate a multi-layer OLED stack, following the technique mentioned above implies studying each layer transport properties for their main transport type (i.e. for instance, NPB which is a hole transport layer will be studied according to hole transport properties. A bivalent transport layer has to be studied according to both hole and electron transport properties). One way to study these material transport properties within the range of thicknesses used *in situ* is to make single carrier devices. These devices are sandwich-type multi-layer stacks of p-i-p or n-i-n types (see Figure 4.13) generally with contacts as ohmic as possible. The intrinsic material in between the doped layers is the material to be studied. Given that before starting the fit procedure, the only indications given on the values to be extracted are rough orders of magnitude, preventing the user from using physically meaningless values, we might have a glance at the challenge of parameter extraction accuracy with a single sample. Hopefully, multiplying operating conditions and thicknesses is a good way to ensure heavy constraints on allowed material parameter sets.

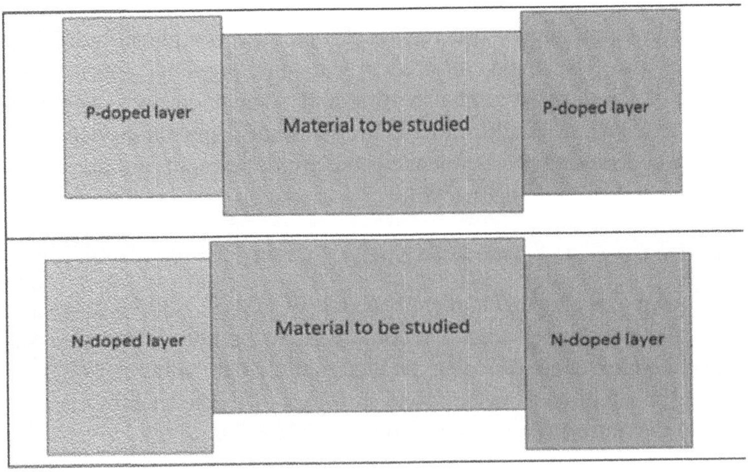

Figure 4.13. *PiP and NiN device structure schemes*

Figure 4.14 illustrates this method with the study of Alq3. A typical single carrier device made up of the following structure: Anode/n-doped layer/Alq3 (X nm)/n-doped layer/cathode was fabricated and electrically characterized at 293, 313 and 333 K. Material parameters extraction was performed through the fit of the several *I–V* curves referring to different material thicknesses (20–70 nm by step of 10 nm) and for different operating temperatures (293, 313 and 333 K), using the ECDM model for electron mobility. We can see that all dependencies are respected with high constraints on the frame of transport behavior imposed by the numerous experimental *I–V* curves. This step should then be repeated for every host material of the multi-layer OLED stack. The ultimate step to validate all the values extracted is their compilation in the final stack and the obtaining of a correct fit of OLED device *I–V* characteristics with minimum changes to these values.

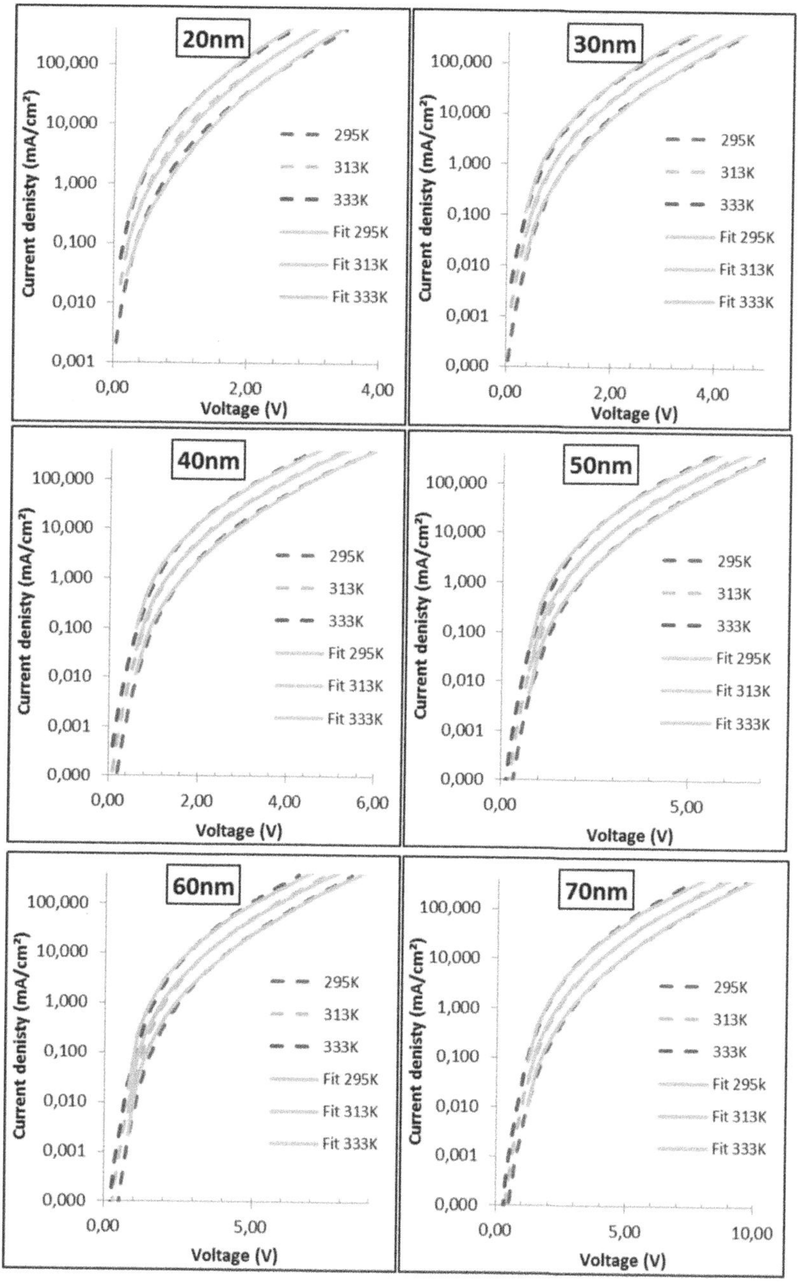

Figure 4.14. *Alq3 single carrier device J–V curves fitted with ECDM mobility model*

4.3.3. *Full OLED stack simulation: example and analysis*

Let's resume from previous steps, assuming the extraction of each layer transport parameters has been completed. The objective is now to simulate the complete structure displayed in Figure 4.15.

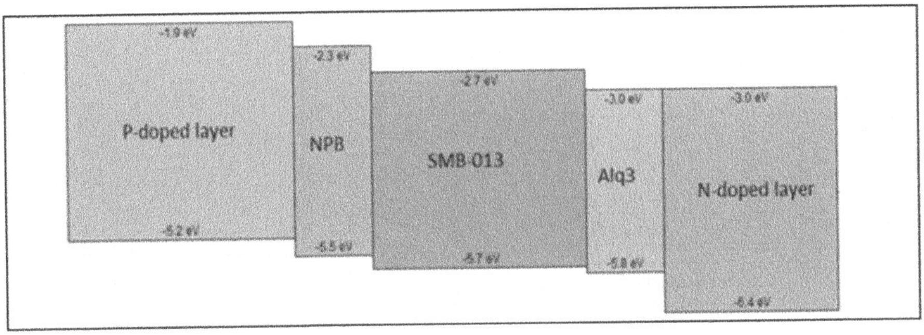

Figure 4.15. *Complete dual emitter white-OLED stack example*

A full white-OLED dual emitter stack *J–V* curve fit is proposed in Figure 4.16. The following fit has been realized through the use of Poole–Frenkel mobility model, and all material mobility parameters were extracted following the same model. Simulation included the management of doped region, induced tunnel injection, non-null built in voltage and host traps as well as states generated by color dopants. The region experimentally described before 2,4 V, caused by leakage current of the device, should not be taken into account as significant in the description of the device transport properties. The regions described for voltages superior to 2,4 V standing for injection regime, trap limited regime and SCLC region are well described by the simulation as can be seen in Figure 4.16.

In order to further investigate the transport properties and the quality of the simulation, we can take a look into the electric field profile across the OLED stack. Figure 4.17 displays such a profile at a given voltage of 5 V (typical working regime for this given stack). Two particular points can be highlighted. Electric field within both n and p-doped layers is negligible when device is emitting, meaning no voltage drop within these specific layers. Most of the field is located in NPB, SMB013 and Alq3, where the recombination takes place. These layers being the least conductive ones in this OLED, it appears logical to witness most of the voltage drop occurring in their specific regions.

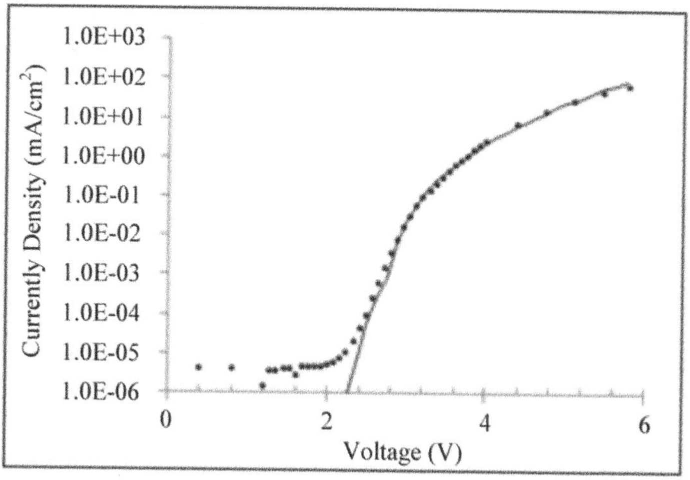

Figure 4.16. *Experimental (dots) and simulated (full line; obtained with Poole–Frenkel mobility model) current density versus voltage curves*

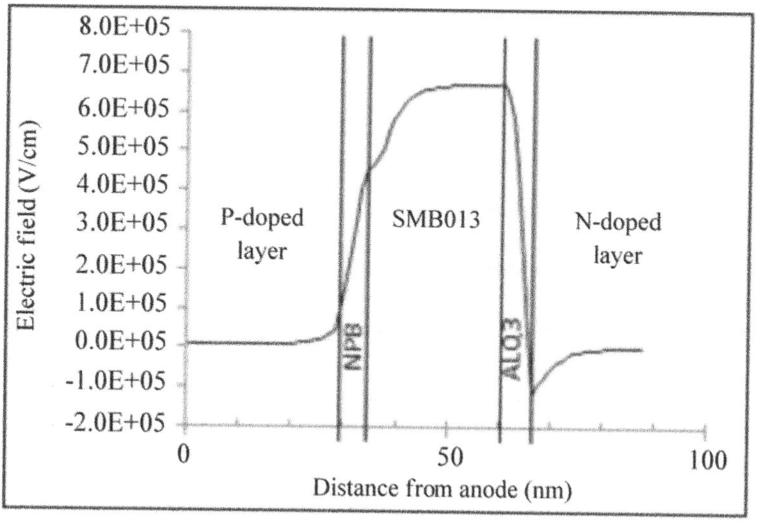

Figure 4.17. *Electric field profile at 5 V*

4.3.4. *Analysis example*

The free electron/hole profile of the previous dual emitter white-OLED structure is displayed in Figure 4.18. We can see the deficit in electrons around the recombination zone. This could let the stack designer think the stack is potentially sub-optimal performance-wise. Indeed, theoretically, a better overall efficiency is reached when the charge balance is reached. Moreover, displaying an excess of holes, in structures that are designed to restrain them past a point, can be detrimental to the lifetime of the device and the performance stability over time. [AZI 99]

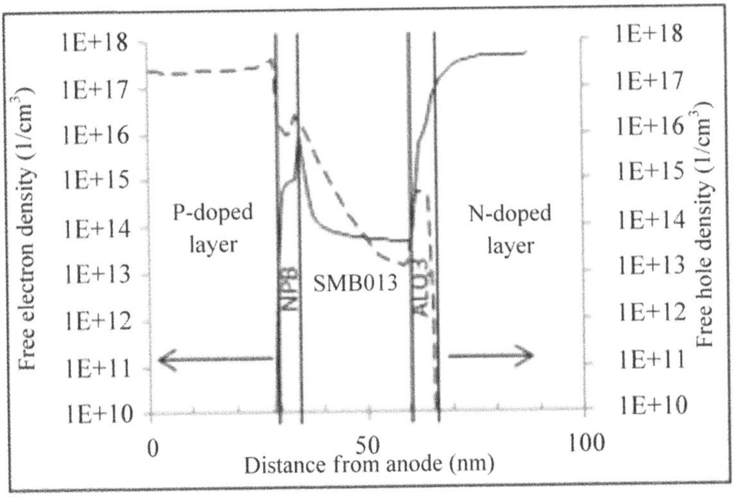

Figure 4.18. *Free electron (full line) and free hole (dashed line) distribution profiles at 5 V*

4.4. Microdisplay simulation limitations

4.4.1. *Electrical/optical crosstalk simulation*

Commercial simulation environments and most lab-made simulation tools using the continuum approach can compute the electrical and/or optical outputs for a given multilayer OLED stack in 1D. However, as far as the authors know, none of these tools can depict or take into account the crosstalk effect, should it be the electrical one or the optical one. By definition, this effect of non-desired electrical stimulation or optical leakage of/in adjacent pixels is a spatial phenomenon that cannot be computed with a 1D tool. Electrical crosstalk is illustrated in Figure 4.19, whereas optical crosstalk is schematically illustrated in Figure 4.20. As depicted, electrical crosstalk can happen when the first injection layer on the pixelated electrode is too conductive and too thick, both at the same time, then allowing current transport in the adjacent pixel. Quantification of these criteria depends heavily on the material used for injection, OLED structure and of course pixel pitch and inter-pixel spacing.

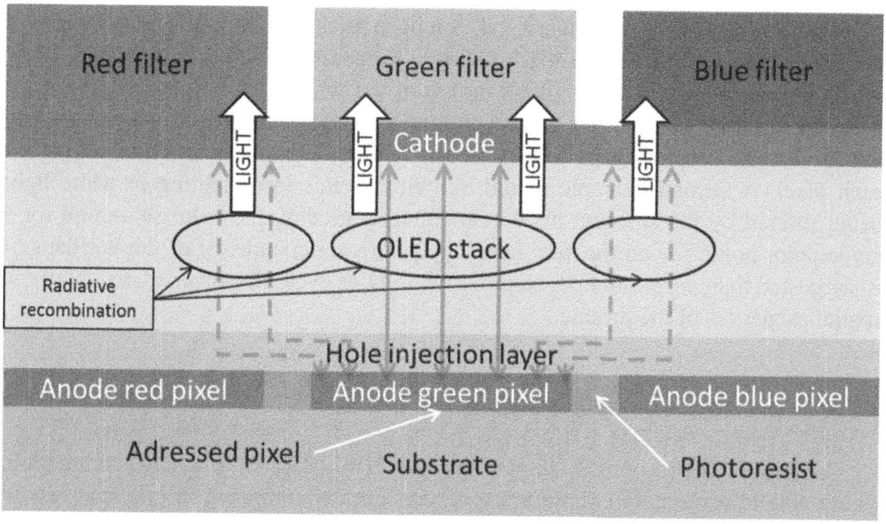

Figure 4.19. *Electrical crosstalk schematically represented in an OLED stack. Full red lines correspond to current path intended by design and dashed lines to current path causing the crosstalk*

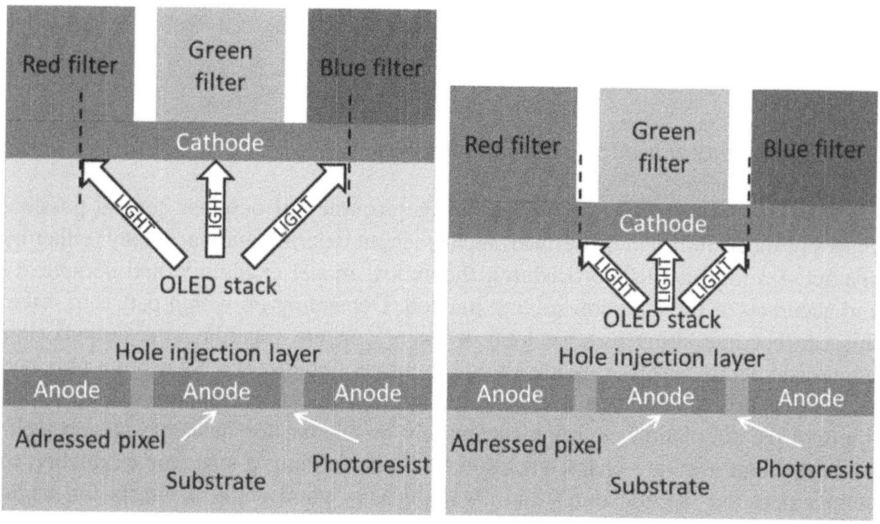

Figure 4.20. *Optical crosstalk schematically represented in OLED stack of two different thicknesses, thickest one represented on the left*

Optical crosstalk, on the other hand, is a phenomenon that occurs on the edges of the pixels, due to several potential reasons. Two examples will be given here. First of all, when placing the color filters on top of the white-OLED microdisplay, gaps between different color filters can be created. Therefore, unfiltered white light can happen to pass through gaps between color filters, resulting in a white edging on each pixel. A second example would be oblique emission resulting in white light being filtered by the adjacent pixel color filter, degrading the color saturation for a given color point. Given the very heavy geometrical dependence of these effects, it is suggested than we should develop our own 2D or even 3D solver to match all the spatial properties of the product.

4.4.2. *Combined electro-optical outputs*

Combined electro-optical outputs are the "Holy Grail" of OLED simulation. Being able to accurately couple transport and optical computation grants access to electro-optical entities, such as current efficiency, which are very important for microdisplays, since current efficiency is related to lifetime performance. To the authors' knowledge, tools capable of doing both types of computation simultaneously do exist. Nevertheless, the use of separated approaches would nowadays still be preferred even though the risk to converge on incorrect conclusions. A critical look at their work is left to the responsibility of the designer. A few lab-made electro-optical simulation tools seem to give interesting added value to simulations as seen in [CHA 12].

4.4.3. *Limitations of accuracy for microdisplays*

The last problem to face is related to the accuracy of outputs. Several possible causes of incertitude can be defined, such as: input determination accuracy (either by characterization or by fit procedure), theoretical model exactitude and complexity and accuracy of the equation solving method. Depending on which part is at stake, this can become either an anecdotic or major concern regarding the outputs. For microdisplays for instance, being able to accurately predict the color point obtained with a certain stack coupled with a certain microcavity is crucial. A few actions can improve the simulation environment, mainly on inputs; for instance: having well-calibrated and accurate characterization tools, performing a systemic evaluation to complement the fit procedures and obtaining very good ohmic contacts for single carrier devices to be as de-correlated as possible from injection limitations.

4.5. Bibliography

[AZI 99] AZIZ H., POPOVIC ZORAN D., NAN-XING HU, *et al.*, "Degradation mechanism of small molecule-based organic light-emitting devices", *Science*, vol. 283, no. 5409, pp. 1900–1902, 1999.

[BUL 98] BULOVIĆ V., KHALFIN V.B., GU G., *et al.*, "Weak micro cavity effects in organic light-emitting devices", *Physical Review B*, vol. 58, no. 7, p. 3730, 1998.

[CHA 12] CHANG Y.-T., LIU S.-W., CHEN C.-T., *et al.*, "A new model for optimization of organic light-emitting device by concurrent incorporation of electrical and optical simulations", *Journal of Applied Physics*, vol. 112, no. 8, p. 084507, 2012.

[PAS 05] PASVEER W.F., COTTAAR J., TANASE C., *et al.*, "Unified description of charge-carrier mobilities in disordered semiconducting polymers", *Physical review letters*, vol. 94, no. 20, p. 206601, 2005

[TOK 99] TOKITO S., TSUTSUI T., TAGA, "Yasunori. Micro cavity organic light-emitting diodes for strongly directed pure red, green, and blue missions", *Journal of applied physics*, vol. 86, no. 5, pp. 2407–2411, 1999.

[VAN 10] VAN MENSFOORT S.L.M., SHABRO V., DE VRIES R.J., *et al.*, "Hole transport in the organic small molecule material α-NPD: evidence for the presence of correlated disorder", *Journal of Applied Physics*, vol. 107, no. 11, p. 113710, 2010.

[YEH 05] YEH P., *Optical Waves in Layered Media*, Wiley, vol. 115, 2005.

5

Addressing OLED Microdisplays

This chapter aims to explain the addressing technologies of organic light-emitting diode (OLED) displays: how images are written on the display, and how the addressing matrices are designed.

At present, two addressing technologies are used for conventional OLED displays: passive matrix and active matrix. Because of its low power consumption and longer lifetime, active matrix organic light-emitting diode (AMOLED) technology is being more and more applied as an improvement or replacement of passive matrix OLED. Passive addressed OLED display is mainly used for small size and low-resolution applications because of their requirements of high-level current with short duty ratio. Conversely, active matrix OLED is devoted to high-resolution display because the whole frame time is exploited for display. Today, conventional size AMOLED displays use Thin-Film Transistor (TFT) technology, which has already been used successfully for display applications, such as flat panel liquid crystal display (LCD) and liquid crystal light valves for projectors.

The case of microdisplays:

Since they were invented, microdisplays of any kind have always been of active matrix type (see Chapter 1). Among the reasons for this we can mention:

1) the question of pixel size: the need for very small pixels imposes semiconductor technologies, then if we pay for the CMOS we have the active matrix;

2) the question of pixel count: the need for a very small resolution on a microdisplay, which needs optical magnification, does not seem relevant application-wise.

Chapter written by Philippe LEROY.

In this chapter, we will first explain the addressing technologies for OLED displays. We will start briefly with the passive matrix technique. Even if it is not used on microdisplays, it will help us to understand the need and functionality of the active matrix one. For this passive addressing, we will explain the classical multiplexing and put emphasis on its limitations. Active matrix addressing will then be presented, and the major electrical as well as the technological constraints will be highlighted. We will then come to the description on how the AMOLED pixel is designed and operates, including in the case of OLED microdisplays.

5.1. Passive matrix OLED display

The passive matrix OLED display shown in Figure 5.1, is the simpler of the two designs, and therefore predestined for use in lower cost applications. It consists of an OLED pixel array that is interconnected via overlapping electrodes. During manufacture, a special process is used to impart a lattice structure onto the patterned ITO anode. Next, the organic substrate and cathode metal are deposited to "automatically" yield the OLED display surface. The intersection of both ITO and cathode defines the OLED pixel size. The "A" cross-section of Figure 5.1 shows the separation element, also called "pillar", acting as a cathode separator in order to achieve the desired isolation of the cathodes.

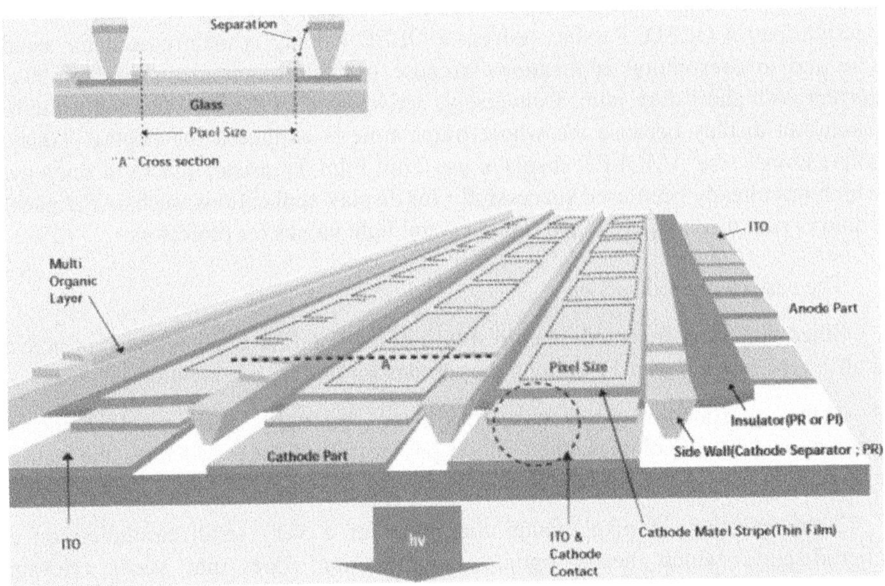

Figure 5.1. *Passive matrix OLED display. This, and all other figures in this book, is available in full color at www.iste.co.uk/templier/OLED.zip*

During the operation of the passive matrix OLED display, individual picture elements are illuminated by applying data current (parallel loading) to the corresponding columns and rows of the matrix. To drive a passive matrix OLED display, electrical current is passed through selected pixels by applying a drive current to the corresponding rows and columns from drivers attached to each row and column. An external controller circuit provides the necessary input power, video data signal and multiplex switches. Data signal is generally supplied to the column lines and synchronized to the scanning of the row lines. Figure 5.2 shows a passive matrix OLED display, where a multiplex switch component (chip-on-flex) is attached to the left side of the OLED matrix, and the data signal's silicon chip supplying the ITO column lines is bonded on the top of the display.

An image output is thus displayed on the panel by scanning through all the rows successively in a frame time. When the bias is on a pixel, it must be it up at an intensity high enough so the desired average brightness is obtained. Therefore, by considering a 64 row panel, the actual emission of each pixel is 64 times higher than the continuous light intensity that is desired.

Figure 5.2. *Passive matrix OLED display with chip on flex drivers*

On the other hand, in a passive matrix OLED display, each line of the picture is addressed sequentially and flashes brightly for a very short time. At the same time, all non-addressed pixels are kept at zero or reverse bias. By scanning the lines sufficiently fast (generating at least 60 pictures per second), the eye integrates all the light flashes into a picture, and no flicker is observed as in a traditional CRT-based TV. As a result, it is a very good property of the passive matrix organic light-emitting diode (PMOLED), as this addressing mode does not suffer any blurring effect also called "hold mode" in comparison with LCD.

In a passive matrix architecture, OLEDs are positioned at the intersections of addressed pairs of anode and cathodes as illustrated in Figure 5.3. The image is built up by making each line of the display flash (current pulse = light pulse) sequentially with an overall frame rate of 50–60 pictures a second. This is achieved by forward biasing the particular line of diodes, which is being addressed, while holding all the other lines of diodes in reverse bias. The high current pulses required during the short line addressing time increased linearly with the number of rows required in the display and are proportional to the multiplex ratio. Reflective metal cathodes (aluminum) are connected to the scan driver, and patterned ITO columns are connected to data driver(s). Data driver based on pulse width modulation (PWM) and/or pulse amplitude modulation (PAM) acts as current sources.

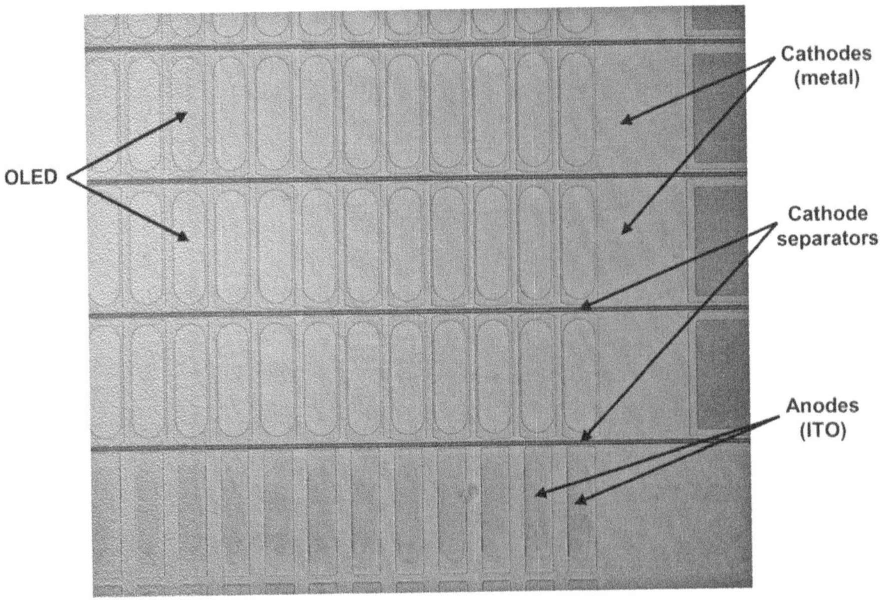

Figure 5.3. *PMOLED pixels*

The equivalent electrical schematic of the PMOLED display is indicated in Figure 5.4. As an example, the last row of this 4 x 4 PMOLED is selected.

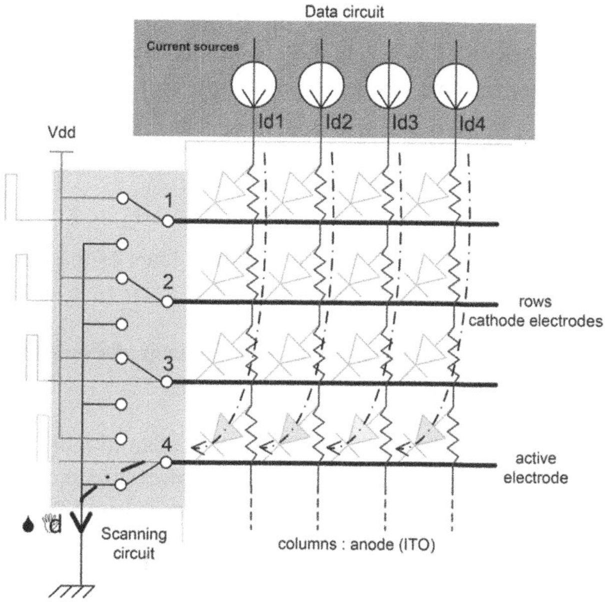

Figure 5.4. *Driving circuit schematic of a passive matrix OLED display*

However, there are problems with image uniformity and power dissipation for passively addressed OLEDs, which are brought about by the need for high current pulses. With reference to the state-of-the-art device, it can be seen that a relatively high voltage (>10 V) is necessary to achieve the high peak intensities of 10,000 cd/m^2 required in a passive addressed display with 100 rows at average 100 cd/m^2 level. The needed current flows through resistive ITO column lines which gives rise to significant voltage drops, causing both non-uniformity in the image and power dissipation. In fact, there are several elements which contribute to the power consumption of a passive matrix display: P_{LED} is a product of operating current I_{LED} and voltage V_{LED} in the LED, and it is reduced when the LED efficiency is increased. There are three sources of power dissipation, P_{LED}, P_{CAP} and P_{RES}:

– Power dissipation in the LED

$$P_{LED} = I_{LED}*V_{LED}$$

P_{LED} is a product of operating current I_{LED} and voltage V_{LED} in the OLED, and it is reduced when the OLED efficiency is increased.

– Capacitive losses

$$P_{COLED} = f*C*V_{LED}^2$$

P_{CAP} is the power required to charge up all the diode capacitances C in the display and can be reduced by lowering the scan frequency f, which can be done, for example, by reducing the frame frequency or multiplex rate.

– Resistive losses (Ohmic losses across resistive row and column conductors)

$$P_{RES} = R*I_{LED}^2$$

P_{RES} is the Joule heating in the row and column lines of the display; it can be reduced by decreasing the resistances R of these lines.

Moreover, with such peak current intensities, the degradation of the EL material is rapid, reducing OLED lifetime.

So, as a summary even PMOLEDs are easy and cheap to fabricate, they are not efficient, and the OLED materials suffer from a lower lifetime due to the high current pulse needed. That is why PMOLED displays are also restricted in resolution and size, as the more lines we have, the more current amplitude we have to provide to keep the desired luminance. PMOLED displays are usually small (up to 3" typically) and are used to display character data or small icons.

5.2. Active matrix OLED displays

5.2.1. *General considerations for active matrix addressing*

In this section, we will review the general consideration for the active matrix addressing of OLED displays. We will compare it with the already very widespread active matrix LCD technology. In both cases, we initially consider the case of conventional small/medium size OLED displays, which are fabricated with TFT technologies on glass. In the following section, we will now focus on the case of OLED microdisplays.

In contrast to the passive-matrix OLED display, an active-matrix OLED has an integrated electronic backplane as substrate and lends itself to high-resolution, high-information-content applications including videos and graphics. This type of display is made possible by the development of polysilicon technology (Poly-Si) which, because of its high carrier mobility, provides TFTs with high-current carrying capability, high switching speed and integrated peripheral circuitry.

There are several key advantages in active matrix OLED displays:

– low voltage and power consumption;

– high resolution;

– large area;

– robust pixel design;

– integrated drivers.

OLED disadvantages

Like any kind of displays, OLEDs are not perfect. The most important drawback is the fact that it costs more to produce an OLED than it does to produce an LCD: between 20 and 30% move expensive at equivalent size and DPI. Of course, the manufacturers, which are strongly involved in AMOLED panels fabrication (Samsung, LGE, AU Optronics, etc), ensure that this should hopefully change in the future, as OLEDs have a potential to be even cheaper than LCDs because of their simple design. OLED does not need any backlight system.

OLEDs have a limited lifetime, which was quite a problem a few years ago. Namely, differential aging between green and red emitting materials versus blue OLED. But, there has been constant progress, and today at least for green and red materials, this is almost a non-issue. Moreover, most recent OLEDs on the market exhibit lifetime capability enough to be used in mobile devices (smartphone and tablets) and probably in TVs. OLEDs can also be problematic in direct sunlight, because of their emissive nature. Newer AMOLEDs, such as Samsung's *Super AMOLED* and *Super AMOLED Plus*, are not very good in that respect. Like its LCD counterpart, OLED also remains very bad under high sunlight conditions.

Before going into move detail about AMOLED technology, it seems interesting to provide a description of active matrix liquid crystal display, which remains the leading display technology. OLED is self-emissive display, and LCD is a transmissive display working like an optical valve that requires an additional backlight system. But, both display systems share the same elements, such as source and gate drivers to build up an image.

Figure 5.5 shows a TFT-LCDs having a sandwich-like structure with liquid crystal filled between two transparent glass substrates. It contains a backlight unit, driving circuitry (source and gate drivers) connected to the pixel array. The TFT-LCD panel of the AMLCD is scanned sequentially line-by-line from top to bottom. Each line is selected by applying the suitable pulse depending on the TFT technology (a:Si or p-Si), which turns on the TFTs in that specific row. At the same time, the data signals are applied from the source drivers to all data electrodes. The voltage applied from the source driver also called "grayscale levels" decides the luminance of the pixels. The storage capacitor Cs maintains the luminance of the pixel until the next frame signal voltage is applied.

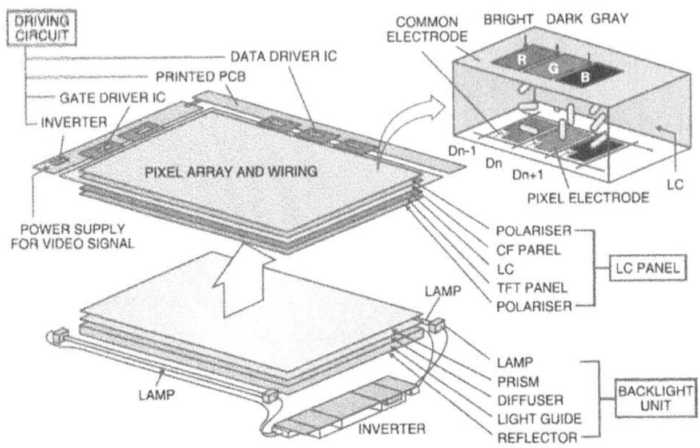

Figure 5.5. *Structure of a standard TFT-LCD module*

Now, the backlight unit only contains an LED which is very efficient and does not require any high voltage inverter circuitry. Cold cathode fluorescent light (CCFL) is largely replaced by LED, and CCFL backlight technology will probably disappear soon.

The TFT panel of an active matrix OLED display works in the same way as for AMLCD. Each individual pixel can be addressed independently via the associated TFTs and capacitors in the electronic backplane. In principle, each pixel element can be selected to stay "on" during the entire frame time. Since OLED is an emissive device, the display aperture factor is less critical, unlike LCD displays where light must pass through an aperture. Moreover, as most of the AMOLED panels available on the market are based on top emission structure, the aperture is no longer an issue. Surprisingly, as we will see further in the chapter, manufacturers do not exploit the opportunity to have larger aperture ratio by using this top emission structure. Figure 5.6 gives an aspect of pixel circuit differences that exist between AMOLED and AMLCD. The main difference lies in the fact that the AMOLED pixel circuit includes an additional TFT (T2) also called the drive TFT that acts as a current generator to provide the requested current to the OLED. Moreover, the AMOLED pixel circuit contains in parallel with data column, a power line electrode supplying the current to an entire OLED column. Furthermore, constant-current TFT (T2) for OLED and the necessary scanning circuitry often based on Poly-Si (integrated drivers) can be built directly on the substrate, thus eliminating the need for high-density (and expensive) interconnects and peripheral drivers. Except for the current generator TFT T2, which only devotes to AMOLED, the situation is the same for AMLCD, as high-end LCD panels based on Poly-Si technology also includes peripheral circuits to address LCD pixels.

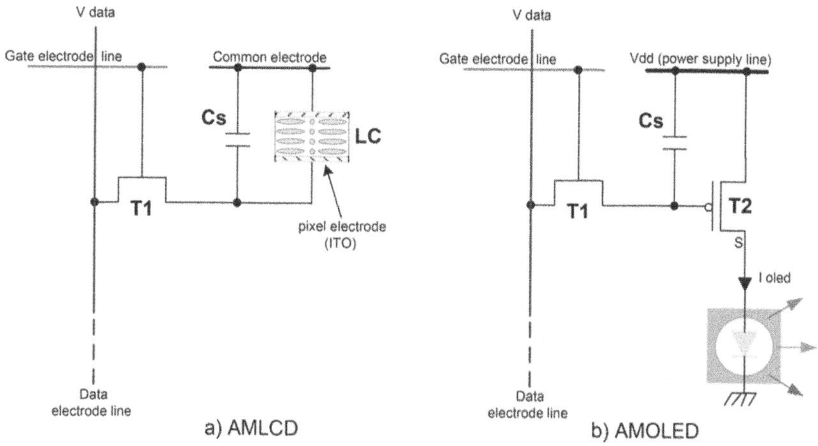

Figure 5.6. *AMLCD a) and AMOLED b) pixel circuit*

Figure 5.7 and Figure 5.8 show through glass substrate (back side view), pixels, TFTs and electrodes layout of active matrix OLED display and its LCD counterpart. Both substrates are based on Poly-Si technology (yellow/orange patterns) allowing peripheral circuits implementation to control gate and data electrodes in order to produce image on the screen. Data electrodes on the top are connected to chip-on-glass (COG). COG acts as a digital to analog converter to provide data voltage to both OLED and LCD pixel circuits.

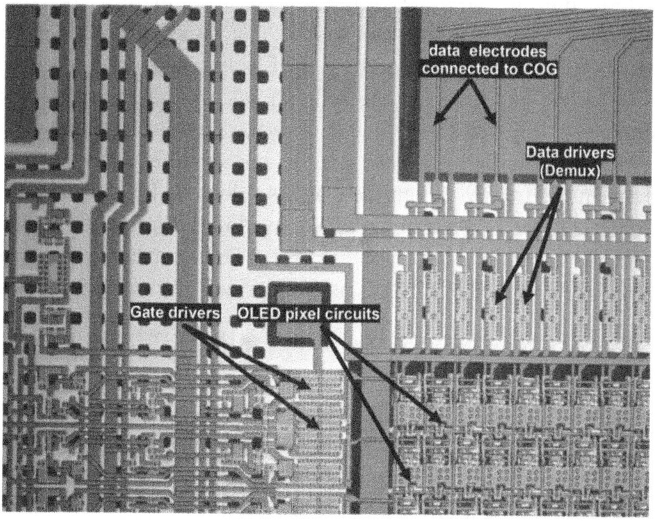

Figure 5.7. *AMOLED TFT substrate (rear view)*

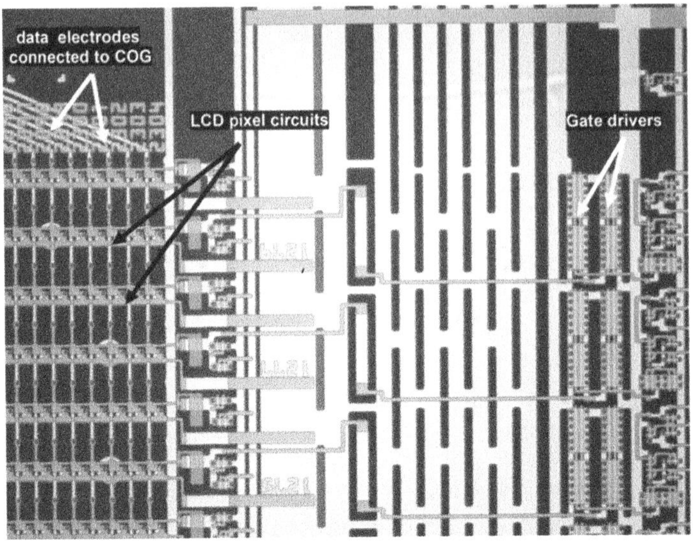

Figure 5.8. *AMLCD TFT substrate (rear view)*

5.2.2. *Two-TFT (2-TFT) pixel circuit*

The goal of the matrix OLED (AMOLED) display is to generate a constant current source at each pixel using TFTs. Each pixel is programmed to provide a constant current during the entire frame time, eliminating the high currents encountered in the passive matrix approach. The simplest circuit for AMOLED displays consists of two TFTs, the address TFT acting like active matrix LCD as a switch and the drive TFT T2. On the other hand, the analog-driving scheme for OLED is similar to the traditional active matrix-driving technique developed for LCDs but with an additional transistor used to convert the analog data voltage into current as show in Figure 5.9. When the pixel is selected, T1 is turned on and V_{dat} is transferred to the gate of T_2. Assuming T_2 is biased in the saturation region, the drain current is given by the gate source voltage of T_2. The data storage V_{data} is stored by C, so that the OLED pixel brightness is maintained during the remainder of the frame.

The pixel structure in Figure 5.10(a) is top emitting, with an *inverted* OLED connected in series with the drive nTFT drain. The pixel array formed by using this inverted OLED structure is common anode connected, while the cathodes are connected separately to individual transistors acting as current sink. However, the ITO anode is deposited last, which provides a more difficult fabrication process causing degradation of OLED performances. As we will see, available AMOLED products do not currently employ inverted OLED structure.

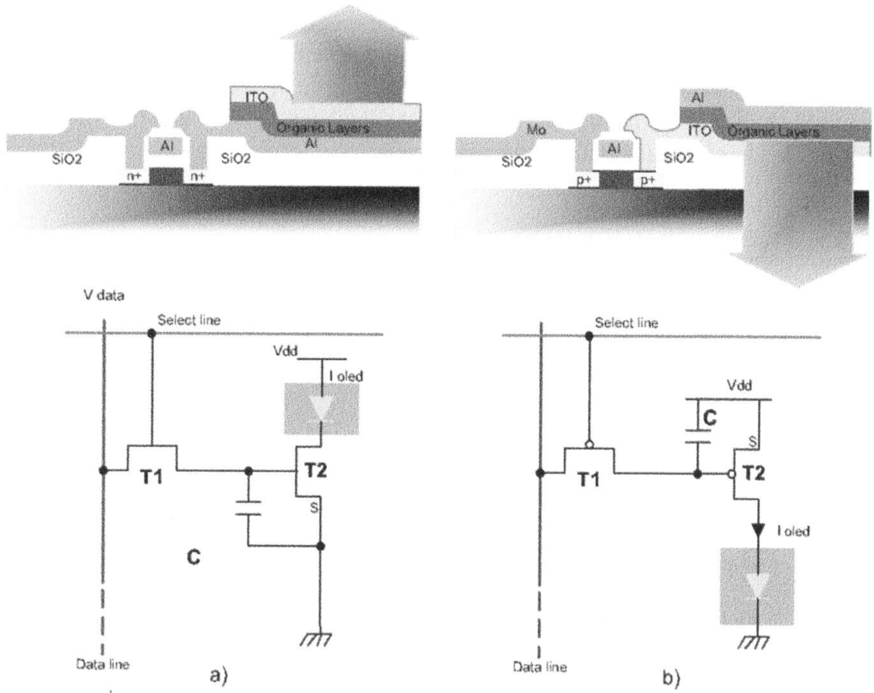

Figure 5.9. *Two possible pixel structures of an AMOLED display*

The pixel structure in Figure 5.9(b) is bottom emitting, with a *regular* OLED connected in series with the drive pTFT drain. This allows the cathode metal to be deposited last and provides a simple pixel cross-section and fabrication process, and makes the OLED current independent of the OLED current-voltage characteristics. Moreover, most of the actual AMOLED displays are based on a top emitting structure in combination with a transparent or semi-transparent cathode and p-TFT drain drive TFT. Unlike the conventional bottom emission structure, where emitted light is partly blocked by the TFT structure, the top emission structure emits light through the opposite side without any interference. In addition to increasing luminance, pixels can be fabricated smaller to realize higher resolution.

The cross-section of the developed top emitting device (Sony) is shown in Figure 5.11. Concerning the top emission structure, the OLED consists of a reflective anode, an emissive layer and a semi-transparent cathode. Light generated in the emissive layer emits outside through the semi-transparent cathode.

The current source pixel circuit in Figure 5.10(b) works well, but requires very high uniformity in the characteristics of T2 across the array, since the current is proportional to $(V_{gs} - V_{th})^2$. Preliminary simulations based on simple calculations, have been made of the current source addressing scheme to determine the effect of the TFT non-uniformity on the current passing through the OLED. Figure 5.11 shows the variation in the driving TFT current calculated for – 10 and 0% variations in Vth. This has been simulated at 16 gray levels.

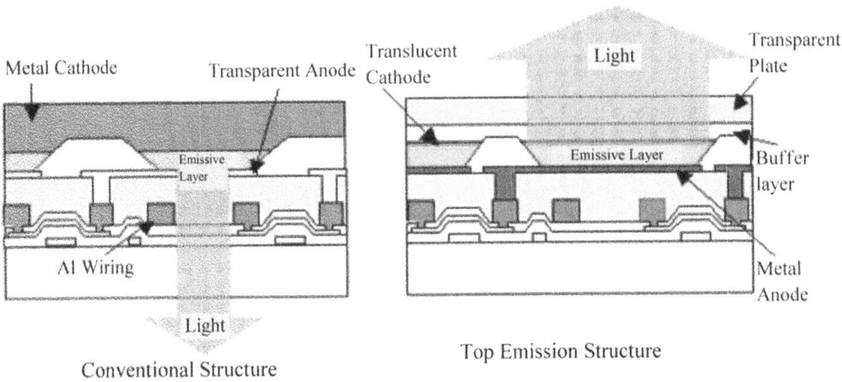

Figure 5.10. *Cross-section of bottom and top emitting device (Sony)*

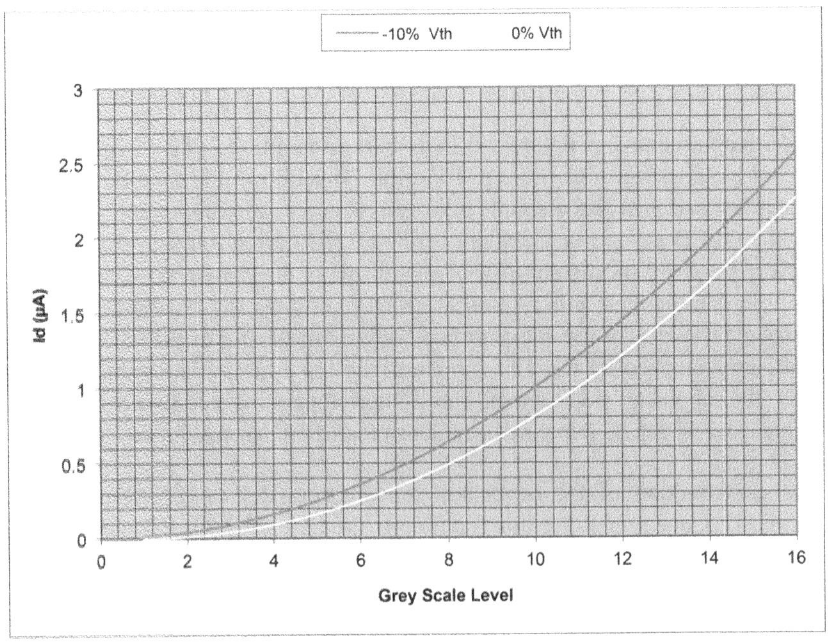

Figure 5.11. *Variation in drive TFT (T2) current for a 10% variation of Vth*

Since OLED brightness is linearly related to the current flow through the device, it is critical to achieve a constant current flow in the OLED to control the display panel brightness uniformity, which can be affected by TFT voltage threshold dispersions.

These luminance variations are tolerable if they occur gradually across a display, because an eye is insensitive to such variation. However, if they occur randomly from point to point, the eye can readily detect these variations. However, it does not occur at random but rather generating some kind of fix pattern noise (FPN), as it will be observed later on in the document.

The problems turn out to be more acute at lower gray levels, as illustrated in Figure 5.13, where the current variation is plotted at a given gray level when Vth changes by 10%.

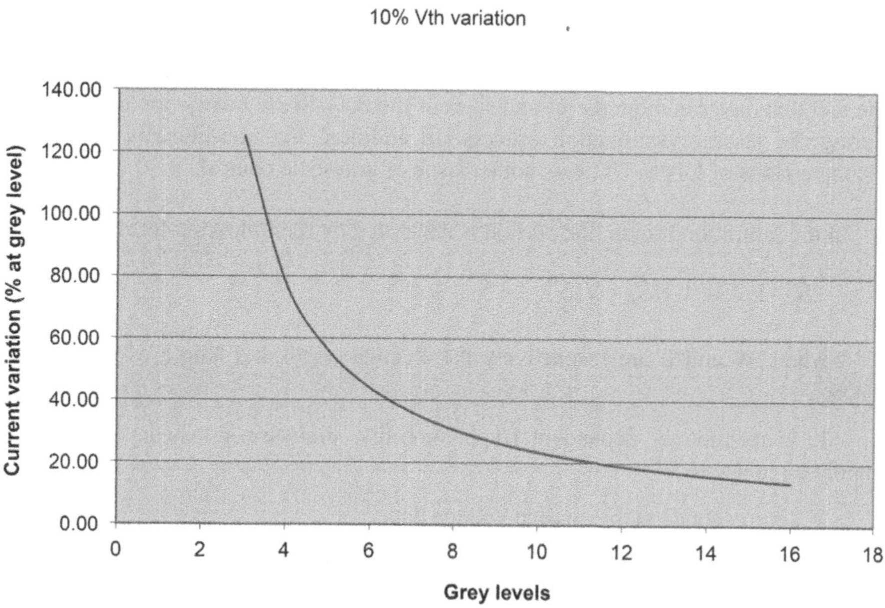

Figure 5.12. *The effect of Vth variation (10%) on driving current*

It can be seen that the current passing through the OLED can vary by as much as 120% for a variation of 10% in the TFT parameters. Any change in Vth results in change of the circuit operating point.

At the lower gray levels, the circuit is operating near the driving transistor's threshold. Any change in threshold results in large variations in current, and therefore brightness for the same signal. Due to the non-linear transfer function, the variation is proportionally smaller at higher current levels.

5.2.3. *Threshold compensation method*

This problem of spatial variation of TFT threshold voltages is currently tackled by using a multi-TFT configuration. Many 4-transistor driving schemes have been proposed in the literature with threshold voltage compensation. The following sections highlight the two interesting approaches to improve the drive-current uniformity of the Poly-Si TFTs applied AMOLED.

The non-uniformity feature of low-temperature Poly-Si (LTPS)-TFTs is due to the fact that the grain boundaries and sizes of the Poly-Si are nearly non-controllable during the laser crystallization process. In addition, the non-linearity of Id-Vgs characteristics of LTPS-TFTs is another issue of grayscale control.

In the saturation region, the current is generated by the following formula:

$$Id = k/2.W/L.(V_{gs} - V_{th})^2$$

– where W and L are respectively the channel width and length of the driving MOS;

– K is technology dependant (type, mobility, dielectric constant, etc) and is expressed in $\mu A/V^2$;

– V_{gs} corresponds to the applied voltage data;

– V_{th} is the threshold voltage, which causes non-uniformity grayscale.

In order to address the ΔVth problem, some kind of built-in compensation circuits are required. Some of the most significant addressing schemes are presented in the following section, including both voltage and current mode data addressing schemes.

5.2.3.1. *Voltage mode data programming scheme*

A new and quite innovative voltage programming addressing scheme based on 4-TFT structure has been recently proposed by the University of Seoul (Korea). Such a configuration, shown in Figure 5.18, which does not employ additional control signals, is very simple.

This voltage-driving pixel structure reported on AM-LCD 02 is based on a 4-TFT and 1-capacitor addressing scheme. This 4-TFT structure is in fact similar to the conventional pixel shown in Figure 5.10(b). T1 is a row line selection switch, C stores sampled data voltage, and T2 drives the current proportional to the voltage stored in C. The T4 and more directly T3 TFTs, play a key role to compensate the threshold voltage variation. T4 and T3 are diode-connected (vgs = vds), and the equivalent schematic is shown in Figure 5.19.

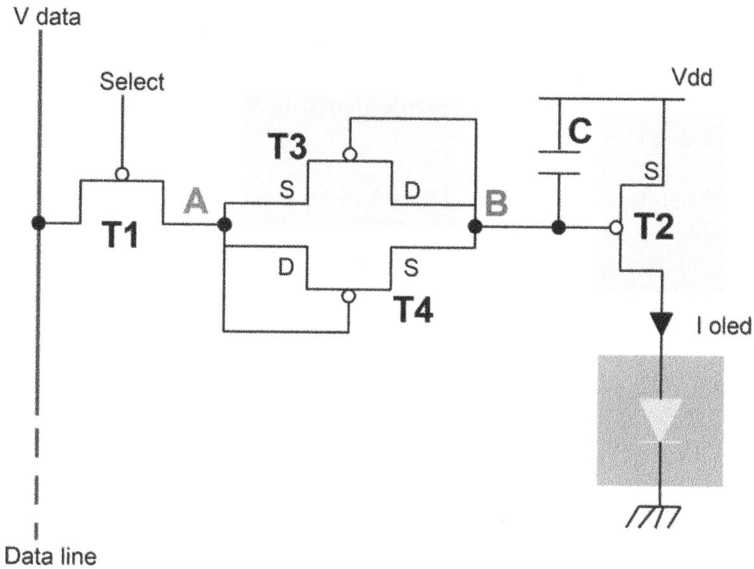

Figure 5.13. *Four-transistor voltage programming scheme*

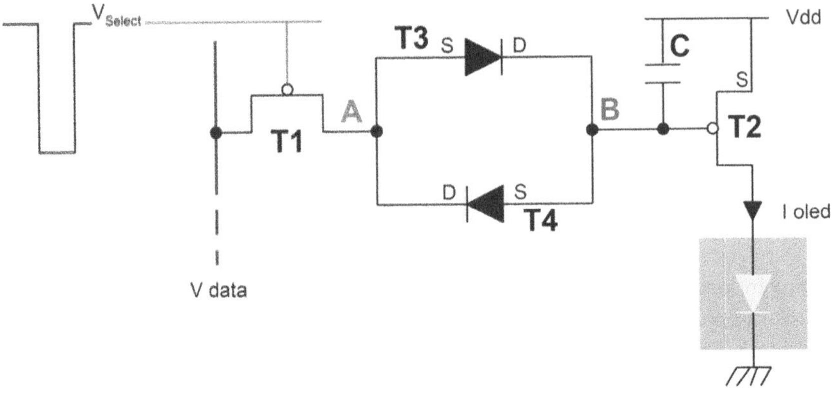

Figure 5.14. *Diode-connected equivalent circuit*

The role of T4 is to reset the previous data storage. When V_A is lower than V_B, T3 turns off and T4 turns on (Figure 5.15(a). Then the reference voltage (reset voltage) of V_{data} modulated by T4 is transferred to C and therefore, the previous voltage stored in C is reset.

When the voltage of the node A ($V_A = V_{data}$) is higher than that of the node B, T3 turns on and T4 turns off (Figure 5.15(b). Then the voltage at the node A is transferred to the node B through T3 until V_B becomes $V_A - |V_{th\_T3}|$ and the modulated data is stored in C. At this time, the drain current of T2 is given by the following equation, in the saturation region.

$$\mathbf{Id} = k/2.W/L.(\mathbf{V_{gs\_T2}} - \mathbf{V_{th\_T2}})^2 \text{ with } \mathbf{V_{gs\_T2}} = V_{dd} - (V_A - V_{th\_T3}) \text{ and } V_A = V_{data}$$

$$= k/2.W/L.(V_{dd} - (V_{data} - V_{th\_T3}) - \mathbf{V_{th\_T2}})^2$$

In such a design, when T2 and T3 are parallel to line beam laser, the threshold voltage of T2 and T3 is assumed to be identical ($V_{th\_T3} = V_{th\_T2}$) then the drain current of T2 becomes: $Id = k/2.W/L.(V_{dd} - V_{data})^2$

Figure 5.15(c) shows the end of the Vth self-compensated operating mode. The drain current of T2 is independent of the threshold voltage of T2 (compensated by the threshold voltage of T3) in such a way that the pixel-to-pixel variation no longer influences OLED current. The way this circuit works is very interesting to understand, because as we will see for off-the-shelf AMOLED displays, this principle of self-detection is widely used.

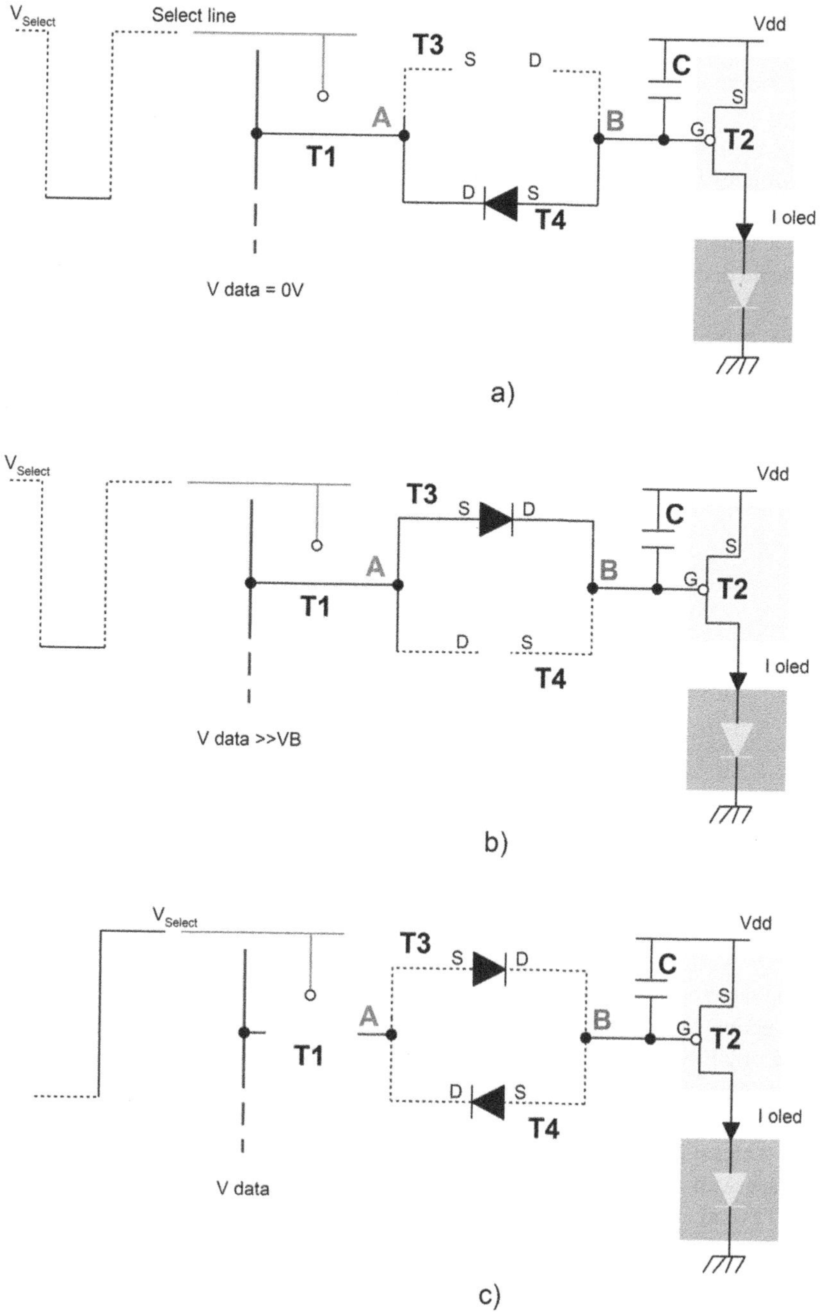

Figure 5.15. *Vth self-compensated pixel operating mode*

5.2.3.2. *Current mode data programming scheme*

As previously said, the conventional 2-transistor driver circuit has an inherent problem of luminance non-uniformity due to the varying characteristics of the individual pixels. The novel pixel addressing scheme (Figure 5.16) developed by Sony, utilizes four transistors, of which two transistors (T1 and T2) are paired up to offset those variations to attain uniform luminance over the entire screen. As a result, luminance non-uniformity can be minimized to facilitate the development of large-scale displays over 10 in.

This current mode data programming scheme can compensate both Vth and mobility variations. Nevertheless, current programming schemes have to manage very small at low brightness, which may lead to unrealistically long programming time.

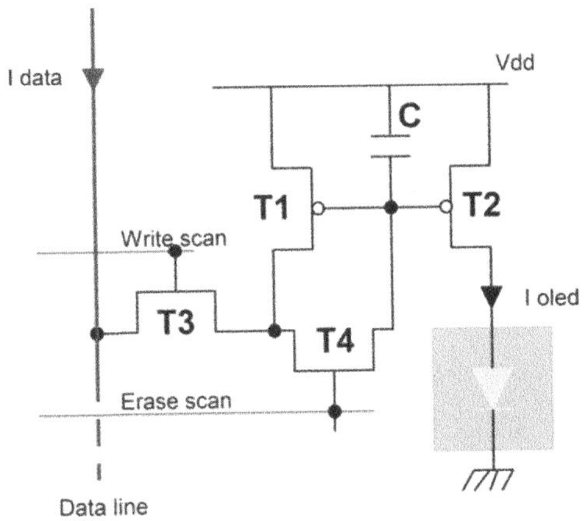

Figure 5.16. *Sony 4T pixel structure*

The developed display features "variable emission duty ration driving" function. By selecting the erase scan while the write scan stays deselected, OLED becomes turned off. Erase scan control lines are separately dedicated to each R, G, and B pixel, which enables easy brightness control.

In the programming operation also called writing mode, both the write scan and the erase scan are selected. Therefore, the data current (I_{data}) is applied to T1 and C with T1 and T2 forming a current mirror, as shown in Figure 5.17(a). At the beginning, T1 and T2 are turned off and Ic = Idata. Then a current starts to flow through T1: $\mathbf{Ic + I_{T1} = I_{data}}$. At least, a voltage signal proportional to Idata is detected and stored in the capacitor C: $\mathbf{I_{T1} = Idata}$. By defining the channel width of T1 as larger than that of T2, Idata becomes larger than Ioled, which makes the writing operation fast enough even at low brightness.

In Figure 5.17(b), both the write scan and the erase scan are deselected, T1 is turned off and I_{oled} is programmed at $\mathbf{I_{oled}}$ **=Idata/4**.

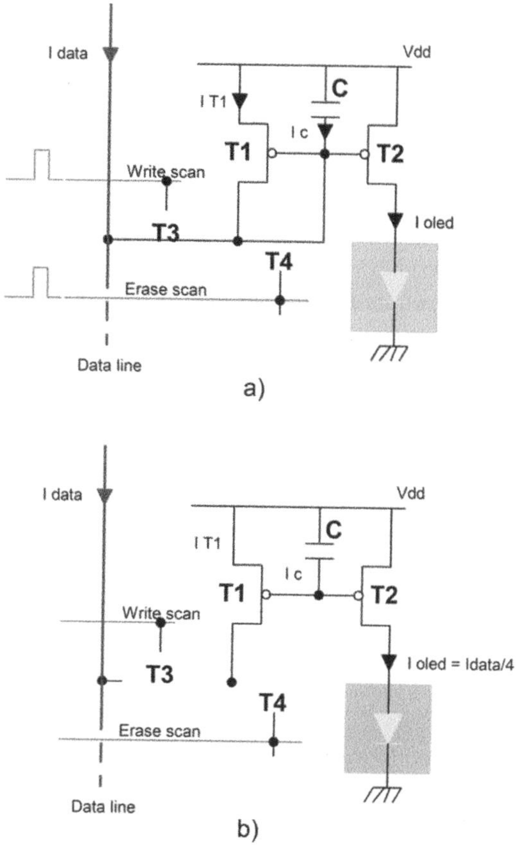

Figure 5.17. *Current-programming pixel circuit*

The two pictures in Figure 5.18 give an aspect of a displayed image between a standard 2-TFT pixel circuit and the current programming 4-TFT pixel circuit. So the compensation method looks relevant, as the kind of FPN existing on conventional 2-TFT circuit has been removed using the 4-TFT pixel circuit. Unfortunately, although nice on paper, such a circuit does not work for larger display because of the parasitic capacitor equivalent circuit of data electrodes that may reach more than 10–20 pF. It means that at a very low current level, this data current spends a lot of time to load parasitic capacitance (actually several frame durations), which is not compatible with the video data rate.

Figure 5.18. *2-TFT pixel circuit on left , 4-TFT pixel circuit right. (Source: Sony)*

5.2.4. *AMOLED pixel circuit and image writing*

5.2.4.1. *Pixel circuit*

Most pixel circuits require adding additional transistors to achieve efficient voltage threshold variation compensation. A lot of AMOLED products (smartphones, tablets, etc.) available on the market use a voltage programming method.

Namely, this AMOLED panel (Figure 5.19) is based on top emission OLED structure. The OLED consists of the reflective anode (OLED layers are deposited onto a reflective metal mirror), the emissive layer and the semi-transparent cathode. Light generated in the emissive layer emits outside through the semi-transparent cathode.

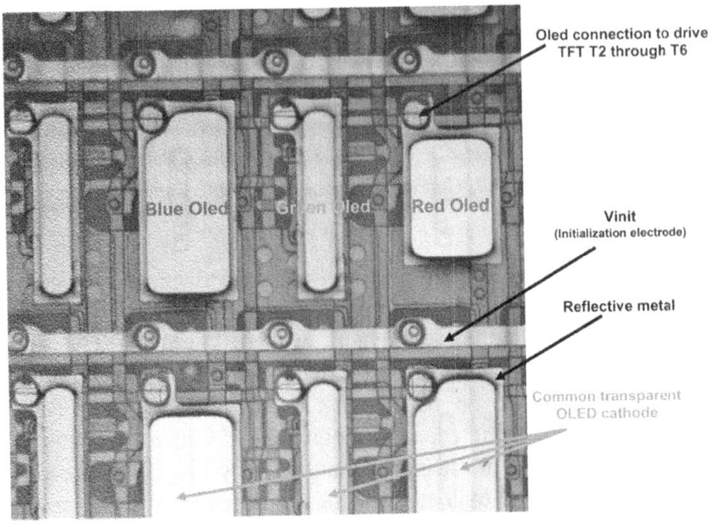

Figure 5.19. *AMOLED pixel circuit (top view)*

Red, blue, green (RGB) OLED areas are not similar, and more particularly blue OLEDs is larger than red and green OLEDs to improve lifetime as well as good colorimetry properties.

The pixel circuit shown in Figure 5.20 is based on LTPS TFT including 6-TFT (T1 to T6). Three TFTs act as single gate switches (T1, T5 and T6) and two other TFTs (T3 and T4) are used as double gate TFTs to avoid leakage currents at the precise nodes of the addressing circuit. The driving TFT T2 is the current generator (voltage-to-current converter) to provide current to the OLED.

– A Capacitor C which is connected to the gate of T2 storing Vdata at frame rate;

– three horizontal control signal electrodes: Init, Scan and Em;

– a data column electrode Vdata;

– a larger power supply electrode Vdd;

– an additional horizontal electrode providing Vinit voltage for pixel circuit initialization purpose;

– a common capacitor electrode;

– a reflective metal element acting as a mirror;

– organic OLED layers;

– a common transparent or semi-transparent OLED cathode.

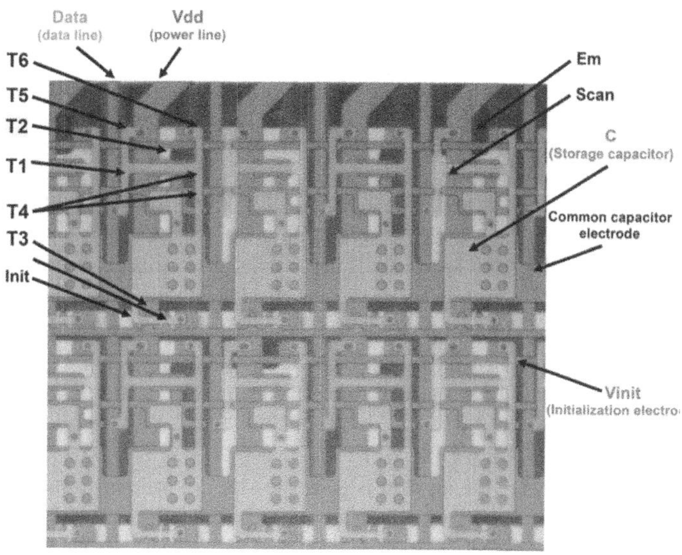

Figure 5.20. *Pixel circuit (rear view)*

Figure 5.21 shows the pixel circuit (6-TFT 1-C circuit) works on voltage programming mode and self-threshold voltage detection principle by the use of a well-known diode-connected structure [LEV 02]:

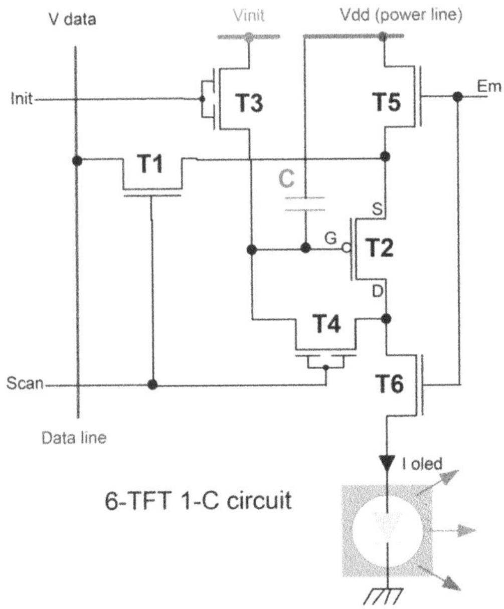

Figure 5.21. *Pixel circuit schematic*

5.2.4.2. *Image writing*

– Initialization phase

T3 (Init), T5 and T6 (En) and T7 (short OLED) are switched on simultaneously. In such a configuration, Vinit is (briefly) applied to the gate of T2 (P type TFT). Assuming this gate voltage is substantially less than Vdd (source electrode of T2), a current flows from Vdd through both T2 and the OLED as shown in Figure 5.27. During this brief initialization phase, a high current amplitude might flow in the OLED. This drawback can easily be circumvented by adding a switch TFT between the OLED anode and the common cathode. Unfortunately, a supplementary control electrode should be requested.

In the saturation region, the current is governed by the following formula:

$$Id = kW/2L(V_{sg} - V_{th})^2 \quad [5.1]$$

– where W and L are respectively the channel width and length of the driving MOS;

– K is technology dependant (type, mobility, dielectric constant, etc) and is expressed in $\mu A/V^2$;

– V_{sg} corresponds to the applied voltage data: V_{sg} = Vdd – Vinit;

– V_{th} is the threshold voltage which causes non-uniformity grayscale.

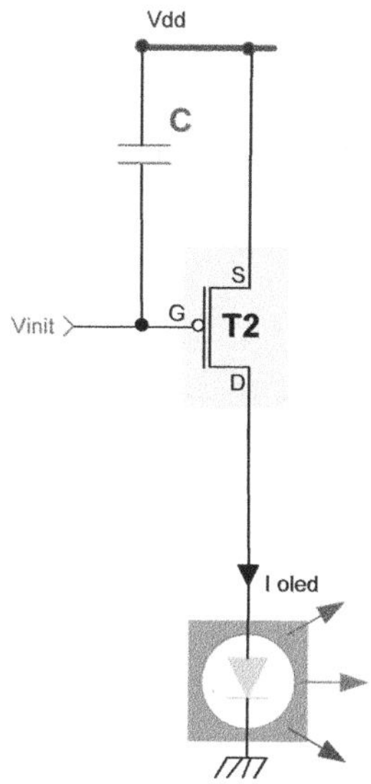

Figure 5.22. *Initialization phase*

– Data writing and threshold detection

This represents the most important and critical phase, as several drawings shown in Figure 5.24 indicate as follows: T3 (Init) is switched off and T1 and T4 turned on. Vdata is applied to the gate of T2, and at the same time drain and gate electrodes of drive TFT T2 are tied together becoming a diode connection structure ($V_{SG} = V_{SD}$ = Vdd – Vdata). The I-V characteristics shown below, are described by the large-signal equation for drain current in saturation (the connection of the gate to the drain guarantees operation in the saturation region):

$$I = Id = kW/2L(V_{sg} - V_{th})^2 \quad [5.2]$$

At this level, the voltage across the storage capacitor **C** is V_C = Vdd – Vdata. Then, both T5 and T6 are switched off by means of control signal En. Therefore, T2 gate voltage is discharged through switch T4 until T2, and OLED is turned off. When T2 turns off, the stored voltage Vc settles from Vdd – Vdata. to **Vdd – (Vdata- V_{SG})**

$$Vc = Vdd - (Vdata.- V_{SG}) \quad [5.3]$$

When T2 turns off, as a consequence **I = Id = 0** and based on equation [5.2], it becomes:

$$kW/2L(V_{sg} - V_{th})^2 = 0 \text{ and finally } Vsg = Vth$$

The threshold voltage Vth has been detected and using [WOO 05] the stored voltage Vc is now:

$$Vc = Vdd - (Vdata.- V_{th}) \quad [5.4]$$

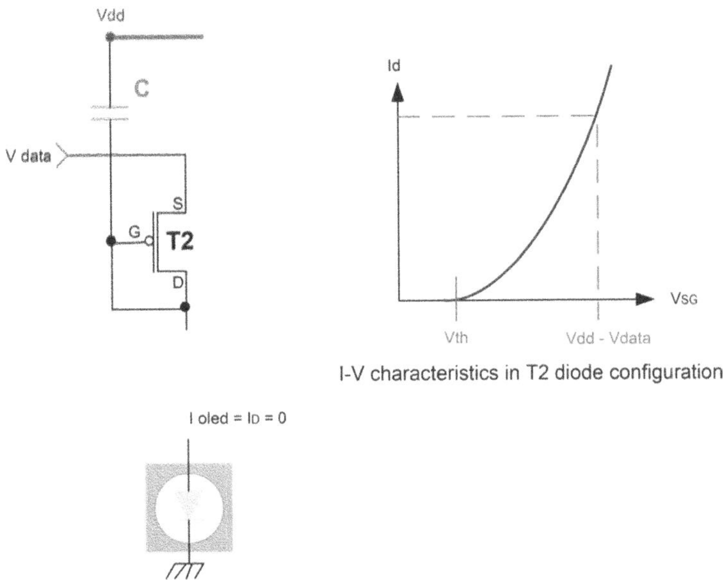

Figure 5.23. *Threshold voltage detection*

– Display (emission period)

Figure 5.25 gives the final result of a standard pixel circuit configuration just after compensation operation. In this period, scan is set to low voltage to block Vdata signal from T1. At the same time, T4 (scan) is also switched off and T2 is no longer diode connected. The storage capacitor C is fully isolated, and Vc still stores Vc = Vdd – (Vdata.-Vth) [5.4]. Shortly after, T5 &T6 (En) are switched on. During this phase, T2 starts to operate in the saturation region and generates a current Id depending on voltage Vc stored in **C**. At this time, the output current of T2 (Ioled), by combining expressions [5.1] and [5.4] and assuming V_{SG} = Vc is given as:

$$\text{Ioled} = \text{Id} = kW/2L(V_{sg} - V_{th})^2$$

$$= kW/2L(Vdd - (Vdata.- V_{th}) - V_{th})^2$$

$$= kW/2L(Vdd - Vdata.+ V_{th} - V_{th})^2$$

$$= kW/2L(Vdd - Vdata)^2$$

Therefore Ioled is independent of threshold voltage.

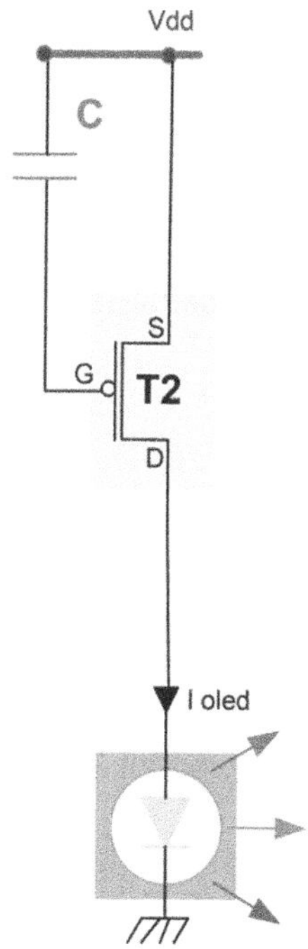

Figure 5.24. *Display period*

5.2.4.3. *Driving waveforms*

Figure 5.26 shows the driving waveforms for pixel circuit:

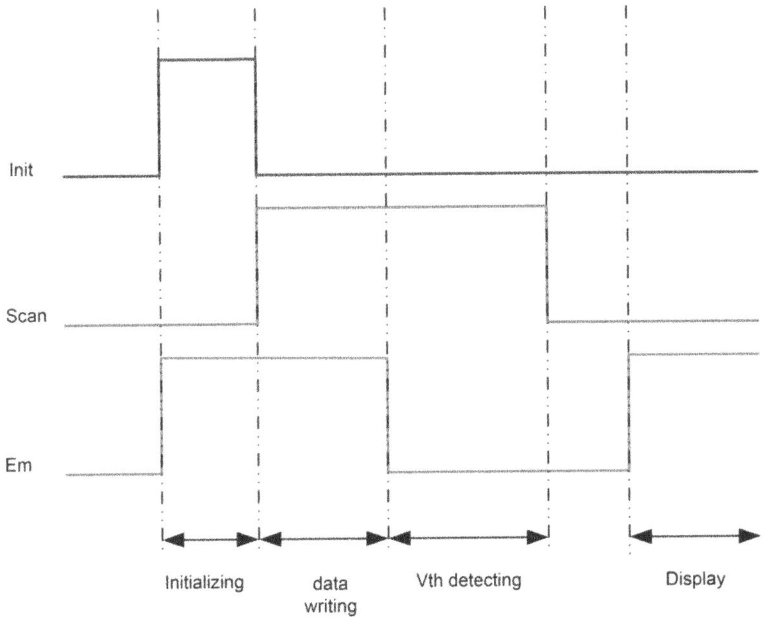

Figure 5.25. *Pixel driving waveform*

5.3. Addressing OLED microdisplays

5.3.1. *Main specificities*

The main difference in the case of OLED microdisplay is that the active matrix is made with a CMOS circuit (Figure 5.26). The color management is also different: due to the small pixel size, for full-color microdisplay application, the only approach is to combine a white emitter with color filters that can be individually patterned (see Chapter 7 for more details on microdisplay fabrication).

Given the obvious power and size constraints for such portable systems, OLED pixels with an area of less than 50 μ^2 are required. Pixel currents are needed from a few hundreds of picoamperes (pA) to tens of nanoamperes (nA), and matching between pixels is necessary to maintain display uniformity.

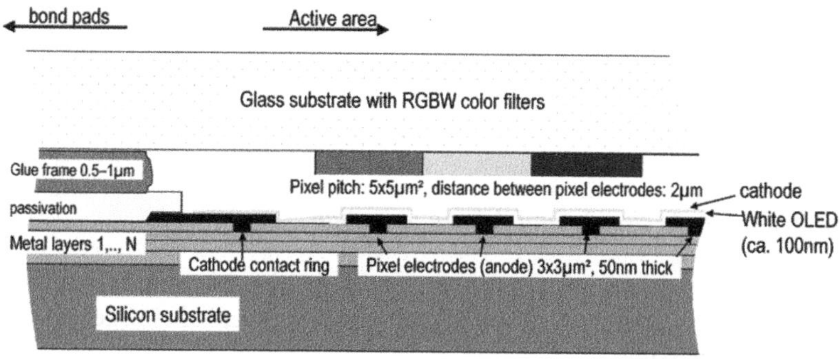

Figure 5.26. *OLED on silicon*

A block diagram of a state-of-the-art microdisplay is shown in Figure 5.27 Bi-directional row and column sequencer circuits are used for addressing individual cells within the pixel array, and internal digital to analog conversion circuits are included for converting the digital input data into the analog signals needed for addressing the pixels. A storage element (capacitor) resides at each pixel cell that is used to set the gray level. In such matrix display architecture, odd columns are driven by data sequencers located at the top of the array and even columns by bottom side sequencers. To obtain a linear grayscale response from the OLED pixels, the digital input data must be formatted with gamma correction. That can be achieved by an on-chip gamma LUT.

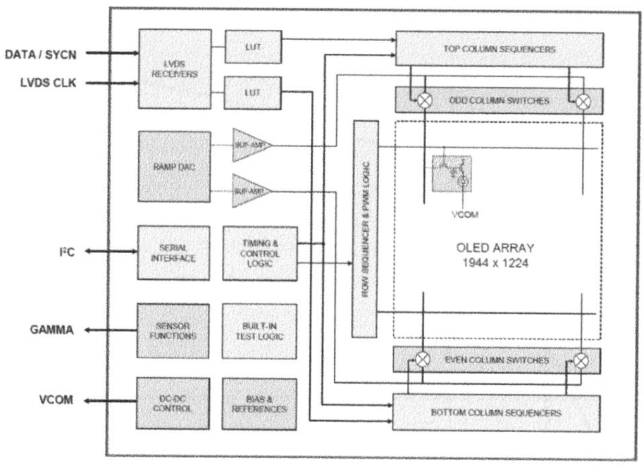

Figure 5.27. *Block diagram of OLED-on-silicon display (source: eMagin)*

Inside the OLED array, a standard 2-TFT + C pixel circuit is indicated. But, as we will see further, the addressing pixel circuit like its Poly-Si counterparts for larger AMOLED displays, is much more complicated to achieve high-quality picture over silicon substrate.

5.3.2. *Pixel electrode circuits and driving operation*

The following section aims at giving several pixel circuit examples that would be suitable to drive ultra-small OLED currents by keeping image quality as homogeneous as possible.

In order to be fully compatible with such small pixels, the current source transistors will be operating in the sub-threshold region. The drain current $\mathbf{I_{DS}}$ in the upper or near sub-threshold region is governed by:

$$I_{DS} = I_{D0}.(W/L).\ e^{[(Vgs-Vt0)/(\eta.k.T/q)]} \qquad [5.5]$$

where:

– I_{D0} is the drain current at threshold normalized for W/L;

– V_{t0} the sub-threshold voltage is explicitly shown;

– η is sub-threshold slope of the process ranging from 1.4 to 1.6 for a typical 0.35 μm process.

The sub-threshold operation of a drive transistor is illustrated in Figure 5.28.

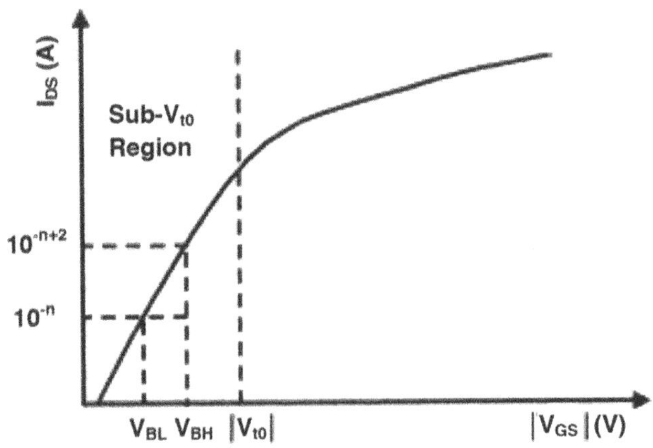

Figure 5.28. *Sub-threshold MOSFET Ids versus Vgs transfer characteristic*

In a general manner, the matching of threshold voltage is a function of area, with Gaussian distribution and standard offset deviation given by $A_{Vt0}/(W.L)^{1/2}$ with a value approximately equal to 10 mV.µm for the process. For illustrative purposes, if a current source transistor is sized with a width equal to 6 m and a 400-nm length, the resulting one-sigma offset is 6.5 mV. The three-sigma case yields 65% error. This high sensitivity to threshold mismatch renders most sub-threshold region mirror designs ineffective. On the other hand, without any threshold matching scheme, the display will exhibit a high level of luminance non-homogeneity: FPN randomly distributed.

Traditional MOS (and TFT) current sources operating in the saturation region are not compatible with such small currents given the area, matching and settling constraints. The issue with such a tiny current (0.8 nA full scale (FS)) is that for current drive we have dramatically slow settling times. Moreover, drive transistors are operated at Vgs << Vth, which is called the sub-threshold region (Figure 5.28), nearly exponential characteristic (1) like bipolar diode. This current is not leakage current, but normal operating current.

In order to address ultra small current driving, *eMagin Corporation* uses the approach to calibrate a high drain current and then shift the Vgs by a calibrated voltage. This is done under the assumption, that Vth is constant in a transistor for all operating regions.

– The first point: the main idea of the pixel circuit presented in Figure 5.29, is to perform a full-scale current calibration using P2, P3 switches and drive transistor P1. Either in the same way as EMAGIN does (see "*An 852x600 Pixel OLED-on-Silicon Color Microdisplay using CMOS Sub threshold-Voltage-Scaling Current drivers*" for detailed explanations), or it could be possible to directly program this full-scale current without sub-threshold shifting. Since this programmed current is the same for all pixels, we could expect in first order (matching) the same errors with respect to settling, sub-optimum guess of pre-charge voltage and leakage currents. Since the programmed current is not changing with time, the charge process is somewhat accelerated. Nevertheless, we could perform the contrast adjustment with this full-scale current also.

– The second point: to do the video level (grayscale video info) with PWM using P4/5/6. Via P6, we put the video voltage on C1. P4 and P5 act as a differential amplifier connected to a current source P1. If Gate (P4) > Gate (P5), the current coming from P1 will flow to GND. Otherwise, it will go through the OLED. (*The ramp waveform drawn should be inverted*). With the ramp waveform, we can program a nonlinear characteristic, etc.

In order to tackle this luminance non-homogeneity, a novel pixel electrode circuit based on calibrated current source operating in the sub-threshold region and connected to a differential amplifier structure is proposed.

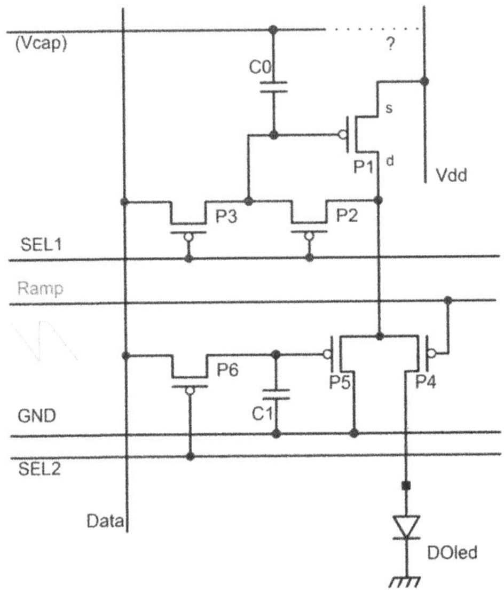

Figure 5.29. *Pixel circuit for microdisplay*

One disadvantage is that the current of the current source is permanently flowing. But, since this current is only a part of the anyhow small total IC dissipation, this maybe not a killing point. There are many options for how long we let the current flow. If we implement the Ramp line as a row line and only put the ramp voltage on the one line (or maybe a packet of n lines) most recently programmed, we get:

– reduced impact of capacitor leakage currents;

– additional current scaling, which brings us away from these tiny currents;

– large area flicker (for bright displays at 50/60 Hz) and potentially motion artifacts.

It is also possible to only activate the complete array during vertical retrace time 1 – 576/625 (pal/secam) or 1-488x/525(NTSC) ~~ 8%.

Regardless of when we let the currents flow, it will always be possible to do the programming when there is no current.

This second section related to the pixel addressing scheme over silicon substrate describes a new method and pixel circuits to drive an OLED microdisplay. Microdisplays are based on a substrate of mono-crystalline silicon, which comprises all electronic circuits. Layers of OLED material are deposited on top of this substrate.

Technical challenges of pixel circuits for microdisplays are on (see also [WOO 05]):

– limited space for circuit elements in available cell area;

– high voltage operation (3.5 V);

– very small OLED diode current (< 1 nA); needing e.g. sub-threshold operation of transistors; sensitivity to leakage currents (see [LEV 02]);

– I/U characteristic of OLED impacting current and thus picture homogeneity;

– dynamic issues of pixel programming due to small currents.

The circuit described in Figure 5.30 all uses the same principle to generate the OLED diode current: by charging and discharging a capacitor at a given clock frequency.

The principle was used before for passive matrix OLED displays (see [SEO 02]). However, the focus here is on circuits for active matrix displays, which allow better display performance and lifetime.

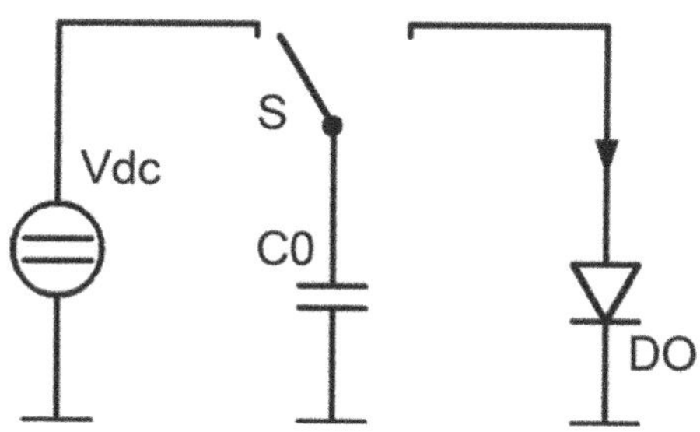

Figure 5.30. *Charge pump circuit*

The charge pump current is given by:

$$I = C * U * f$$

It is proportional to the charge voltage, more correctly: the voltage difference between the charged and the discharged state. Good picture uniformity can be achieved, since frequency, capacitor matching and probably also a good voltage precision can be implemented.

MOS capacitors will have a capacitance per area of around 5 fF per micron square. For a 5 x 5 μ pixel, let us assume, we can use 20 f (2 × 2 μ) for the charge capacitor. Operating the charge pump at 32 kHz and assuming a voltage for the full scale of 2V, the average diode current will have a FS value of Imax = 1.28 nA.

To avoid high peak currents, when switches are closing, a ramp like control voltage is used.

A more detailed analysis is also desirable to check, what the impact of Vth mismatches (threshold voltage of MOS transistor) can be. A first rough analysis has shown that on mono-crystalline silicon, and assuming FS voltage amplitudes of 2V, the Vth mismatch is probably negligible. Anyhow, some mismatch compensation is included in some of the following circuits of active charge pump pixels such as a first version shown in Figure 5.31.

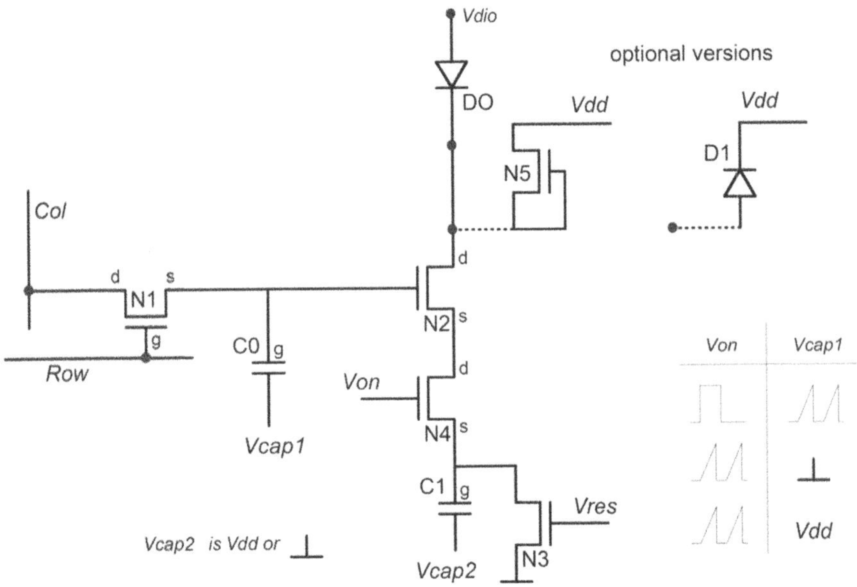

Figure 5.31. *Active charge pump circuit using N channel MOS transistor*

During the programming phase, the C0 (the node g) is set to the signal voltage. This is done by applying the programming voltage to the column line Col and applying a high pulse to the row line Row. The programming is usually done synchronously, line by line, with the incoming video signal. During other times, there is the repetitive charge pump process, in order to generate a current through the load, which corresponds to the programmed voltage. The capacitor C1 is repeatedly charged to (roughly) the voltage on C0 by making N4 conductive for a time. The source follower transistor N2 and the voltage on C0 determine when the charging is completed. The charge current through N2 and N4 is also the current through the load DO. In between, C1 is discharged, being shortened by N3.

Driving Vcap or Von with a voltage of sawtooth shape has a purpose of avoiding high peak currents, but to get a current, which is about constant most of the time.

Controlling Vcap1 can also help to shift the possible voltages on the column Col to a suitable range during the programming period, and afterward to a different level for best charge pump operation.

The optional clamp circuit can comprise a transistor N5, acting as a diode, or a diode D1, if preferable with the process used. The purposes can be different:

– High voltage operation and low voltage CMOS process: in case, the voltage Vdio exceeds the (maximum possible) IC supply voltage Vdd, the diode, e.g. N5, limits the voltage at drain of N2.

– Testability: the diode N5 or D1 can form a current path also in the case, there is not yet an OLED device deposited on top of the silicon chip. Manufacturing tests of raw wafers with reasonable test coverage are possible. (Vdd should comprise separate column or row lines in this case.

– Vth compensation: we could, for example, generate a time period, where the OLED intentionally shortened by the diode by pulling down Vdio. Then, Vcap1 and the gate of N2 would be ramped up slightly, so that N2 just starts to conduct. At this point, C1 has reached a voltage, which is lower by Vth of N2, so that the following true current flow is free of the Vth of N2 and its tolerance.

Vcap 2 (and Vcap1, in case it is a constant voltage), can be fixed to either Vdd or GND. The GND case is preferable to avoid additional AC current components between the supply rails. However, for capacitor implementation reasons, the Vdd case can be preferable. Figure 5.32 above is different from Figure 5.31 in the sense, that N2 and N4 are changed in order. The overall function is the same. One of the two variants can be preferable for layout reasons.

Figure 5.33 below shows a similar topology as Figure 5.32, however, now using p-MOS devices. This is probably the more interesting case, because the most common OLEDs have a common top electrode, which is the cathode (Vdio). Again,

P4 and P2 may be swapped in order. Capacitors have their reference terminals connected to Vdd or GND. Swapping the order of P2 and P4 is also an option here.

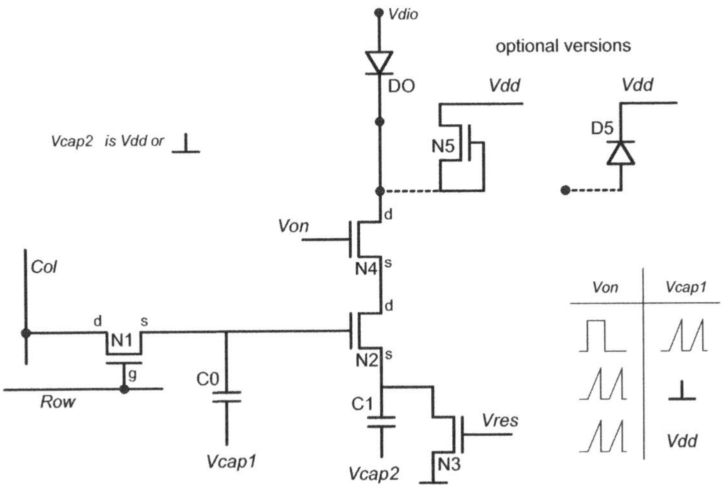

Figure 5.32. *Active charge pump circuit n°2*

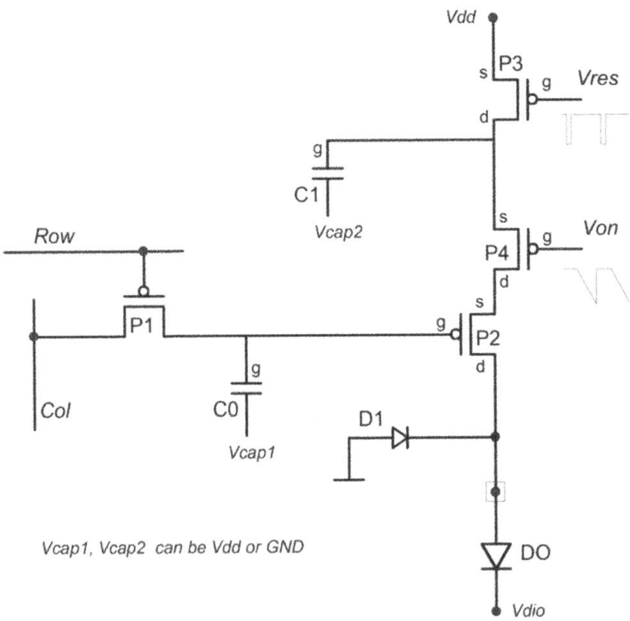

Figure 5.33. *Active charge pump circuit using P-channel MOS*

Figure 5.34 is a simplified circuit omitting the switch transistor N4. When shortening C1 with N3, Vcap must be driven low to disable N2 becoming conductive.

During programming, we either control Vcap in a good way, or we accept that the capacitor C1 is charged to a wrong value the one time we do the programming of C0. Assuming we do three current pulses per lines and we have ~500 lines, the error will be less than 1/1500 (probably acceptable).

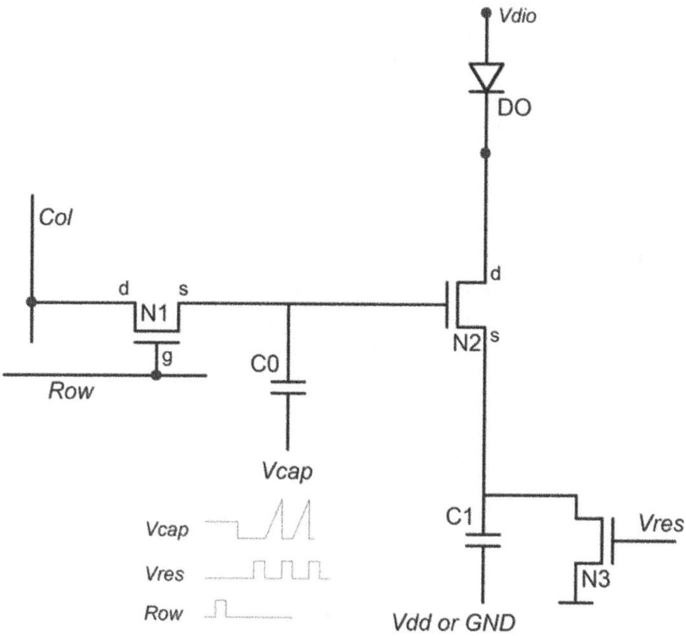

Figure 5.34. *Simplified active charge pump circuit using N-channel MOS*

That completes the description of possible pixel addressing scheme based on charge-pump circuit to provide an efficient and homogeneous method to drive ultra-small OLED over silicon substrate.

5.3.3. *Innovative pixel circuit on silicon backplane*

As indicated in the introduction, the pixel driving current of micro-OLED displays is several nano-amperes and less because the emission current of OLED devices is proportional to the pixel size. Consequently and because of processing rule variations over silicon backplane, the pixel-to-pixel luminescence has non-uniformity due to more less slight variations in the driving current. A pixel circuit is required to precisely control these variations in the driving current. In recent years, many display manufacturers have developed pixel compensation circuits on LTPS or Si crystalline backplane. A standard pixel circuit requires many transistors (sometimes up to seven including additional control signals) to compensate the variations in the driving current. However, it is difficult to produce so many transistors on sub-pixel areas much smaller than 10 um. To overcome this problem, a new pixel circuit has been developed on a silicon backplane [ONO 12], which consists of only two MOSFETs and is capable of compensating variations in both the threshold voltage (Vth) and the mobility (μ) of the driving transistor. Consequently, a smallest pixel size of 3.3 um x 9.9 um has been achieved having a good pixel-to-pixel luminance uniformity. Figure 5.35 shows a schematic diagram of new pixel circuit shows a schematic diagram of the pixel circuit. The pixel circuit uses voltage-programmed current driving method. The pixel circuit consists of two n-channel MOSFETs (T1 and T2), two capacitors (Cs and Coled), a gate line (select line), a switchable power supply line (Vdd) and a data line to provide Vdata. T2 is the OLED driving transistor that controls the emission current of the OLED device. T1 is a switching transistor for voltage-programming operation. Cs is a storage capacitor to hold the voltage of T1 during the frame time. Coled is a load capacitor of the OLED device during writing of signal voltage.

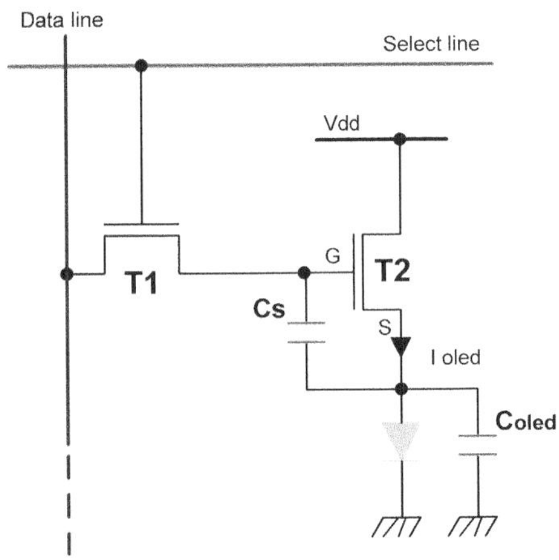

Figure 5.35. *Schematic diagram of new pixel circuit*

In comparison with its Poly-Si counterpart used in most AMOLED smartphones, this addressing scheme is simple, as it is based on the standard 2-TFT structure. The main difference is that a switchable power supply line is employed to select either Vdd (power supply) or Vinit for initialization phase.

– Pixel driving method

Figure 5.36 is a timing diagram of new pixel circuit shows the timing diagram of this new pixel circuit. Vdd controls the light-emitting period. Select line provides the gate pulse of T1 for controlling the write and Vth cancel operations. Additionally, in the new driving method, two data voltages Vref and Vdata are alternately applied to the data line during the horizontal period. Vref is a constant reference voltage for compensating variations in Vth of T2. Vdata is the signal voltage that is correlated with the luminescence of each pixel. The new pixel driving method automatically compensates the electrical characteristics of the driving transistor through two operations, namely, the Vth cancelling operation and also the mobility cancelling operation.

The new pixel driving operation, which is described below, consists of the following four operations:

1) reset operation; 2) *Vth* cancel operation; 3) write and mobility cancelling operation; and 4) light-emitting operation.

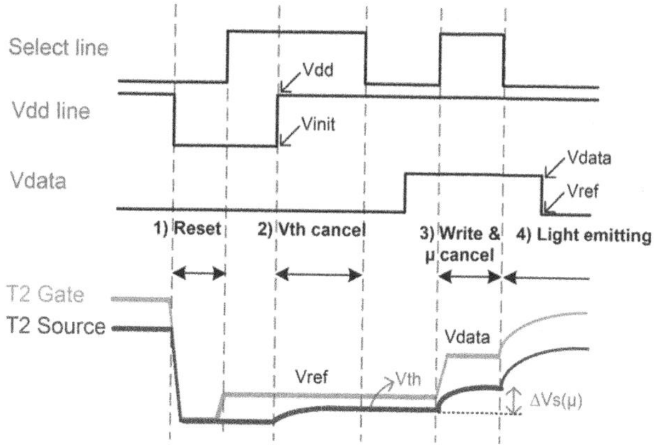

Figure 5.36. *Timing diagram of new pixel circuit*

1) The initialization operation is shown in Figure 5.36, a timing diagram of new pixel circuit. The switchable power supply line provides a positive supply voltage (Vdd) to the initial voltage (Vinit). Vinit is set to a low voltage in order to completely turn off the OLED device.

2) Vth cancel operation is also shown in Figure 5.36, a timing diagram of new pixel circuit. When T1 is turned on, Vref is applied to the gate node of T2 from the data line. Then, the select line changes to Vdd from Vinit, and T2 is turned on in the saturation region. Since the driving current flows to the OLED device, the source node of T2 gradually increases and the driving current decreases. Finally, the source

node reaches **Vref-Vth**, and T2 is then turned off. Consequently, the Vth of T1 is stored in C1 after select line is switched off.

3) Figure 5.36, a timing diagram of new pixel circuit shows the signal data writing and the mobility cancelling operation. Now, data line changes to Vdata from Vref. When T1 is turned on again, *Vsig* is applied to the gate voltage of T2 from the data line and T2 is turned on again. The source voltage increases when select line is high. As a result, the source node increases by ΔVs(μ). Since Vth compensation has previously been accomplished, ΔVs(μ) approximately depends on the mobility of each individual T2 of the silicon matrix. Therefore, the ΔVs(μ) is much greater for higher mobility pixels. Thus, the mobility variations of drive TFT T2 is performed by controlling ΔVs(μ) in this period.

4) Light-emitting operation. When T1 is turned off, the gate-source voltage (Vgs) of T2 is stored in Cs. The value of **Vgs = Vdata – Vref + Vth - ΔVs(μ)**. The emitting current Ioled flows in the OLED. The gate and source voltages of T2 together increase by maintaining the Vgs of T2. Finally, and completing this last phase, the source voltage reaches the emission level. Assuming the drive TFT T2 works in the saturation region, Ioled is given by the following formula:

$$IOled = \mu.Cox.W/2L.(V_{gs} - V_{th})^2 = \mu.Cox.W/2L.(Vdata - Vref + V_{th} - \Delta Vs(\mu) - Vth)^2$$
$$= \mu.Cox.W/2L.(Vdata - Vref - \Delta Vs(\mu))^2$$

– where W and L are respectively the channel width and length of the driving TFT;

– μ is TFT field effect mobility typically around 1000 for mono-crystalline TFT;

– V_{gs} corresponds to the applied voltage data;

– V_{th} is the threshold voltage;

– Cox is TFT channel capacitance per unit of area.

The final equation indicates that both the mobility and the Vth variations of T2 have been suppressed, allowing uniform pixel-to-pixel luminance despite the ultra low size pixel.

5.4. Bibliography

[BOW 10] BOWER C.A, MENARD E., BONAFEDE S., *et al.*, "Active-matrix OLED display backplanes using transfer-printed microscale integrated circuits", *Electronic Components and Technology Conference*, pp. 1339–42, 2010.

[ONO 12] ONOYAMA Y., YAMASHITA J., KITAGAWA H., *et al.*, "0.5-inch XGA micro-OLED display on a silicon backplane with high-definition technologies", *Society for Information Display Conference Proceedings SID'08 Digest*, pp. 950–53, 2012.

[LEV 02] LEVY B.G., EVANS W., EBNER J., *et al.*, "An 852x600 pixel OLED-on-silicon color microdisplay using CMOS subthreshold-voltage-scaling current drivers", *IEEE Journal Of Solid-State Circuits*, vol. 37, no. 12, pp. 1879–89, December 2002.

[SEO 02] SEO J.W., KIM H., KIM B., *et al.*, "A charge-pump active-matrix pixel driving method for organic light-emitting diode display applications", *Japanese Journal of Applied Physics*, vol. 41, pp. 7391–96, 2002.

[WOO 05] WOODBURN R., KELLY G., UNDERWOOD I., *et al.*, "On the driving of OLED micro-displays", *Proc. Of Euro display 2005 Conference*, Edinburgh, Scotland, pp. 408–11, September 2005.

6

OLED Microdisplay Fabrication

The specificity of OLED microdisplays fabrication lies mainly in six points:

– the small size of pixels preventing RGB sub-pixel individual deposition. The deposition of white emitting OLED structures is then mandatory;

– consequently deposition or assembly of color filters that requires a lot of operations onto a fragile device;

– the inter-pixel insulation;

– the use of a non-transparent Si substrate that requires an adaptation of the regular deposition machine tool that is originally made for deposition onto glass substrates;

– the OLED structure must be a 'top-emitting' one;

– low level of acceptable defects, which are related to the size of pixel; a given defect is more detrimental to the microdisplay than to a display bearing pixels of several tens of micrometers.

6.1. Fabrication of CMOS active matrix

6.1.1. *General considerations*

As explained in the introduction section, active matrix for OLED microdisplay is a CMOS circuit. The top surface of such a circuit is an array of electrodes constituting one end of the OLED structure of each pixel, which can be either anode or cathode depending on the circuitry and OLED type (PiN or NiP), and serve as a hole or electron charge injection termination, respectfully.

In general, a complementary metal–oxide–semiconductor (CMOS) circuit is composed, from bottom to top, of the core device levels (active regions with ion

Chapter written by Christophe PRAT, Tony MAINDRON, Rigo HEROLD and François TEMPLIER.

implantation, thin gate insulators, contact metals, etc.), on top of which are the interconnect levels: a stack of four or more metal levels plus insulators, which make the connections between CMOS devices and circuits. Until this stage, the fabrication process is standard. And the associated circuit only performs purely electronic functions (processor, memory, analog functions, etc.). For some circuits, for particular applications, additional layers are added to provide special functions. This is then called an "above-IC"" process. The CMOS circuit is then more than a simple electronic circuit. One of the most known and widespread examples are the CMOS imager, or the microelectromechanical systems (MEMS), etc. Associated process steps of such CMOS circuits are dedicated and specific to the application, and may need process development, non-standard materials, etc. This is the case of CMOS for OLED microdisplays.

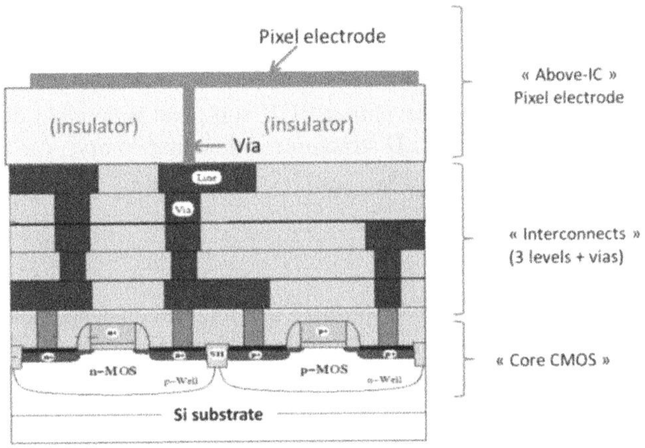

Figure 6.1. *Schematic cross-section of CMOS active matrix for OLED microdisplay, showing separate parts: standard "Core CMOS" and "Interconnects" (in this example: three levels), plus specific "above-IC" pixel electrode. This, and all other figures in this book, is available in color at www.iste.co.uk/templier/OLED.zip*

The specific process needed for the OLED active matrix is the realization of the array of electrodes, together with the associated insulation with underlying layers, and the contact with the bottom circuit. The global structure then looks as shown in Figure 6.1.

6.1.2. *Specificities of the circuit*

The specificities and difficulties of these particular processes mainly lie in:

– type of metal electrode: it concerns OLED device performance. The electrode should combine the highest reflection coefficient to ensure a maximum of OLED light emission upward (we are in top emission configuration), and appropriate work function to match the molecular energy levels of the carrier transport materials in the OLED stack (see Chapter 1 and 2);

– surface quality: due to the very small thickness of the layers present in the OLED stack (a few nanometers), electrode surface must present a very small roughness, to ensure good contact and limit risk of vertical short circuit.

In addition, these materials should have a good ability for being processed with semiconductor manufacturing equipment: they should be deposited using conventional tools (in general physical vapor deposition type: sputtering or evaporation), should be etched at a high rate with high selectivity, etc.

One of the most important performance indicators of these processes is the insulation quality between neighboring electrodes. In fact, a single short circuit between two of them will give rise to pixel defects, visible on the display and might reject the whole CMOS circuit. The trick is that the inter-pixel space might be as small as possible to achieve the best pixel pitch (highest dpi), without insulation defect to guaranty high yield.

The role of interconnect levels: these levels are used for the realization of active matrix lines (select line connecting select transistors), columns (data lines, Vdd lines, etc.), busing, interconnects between transistors and capacitors inside the pixel, the drivers and the fan-out.

6.1.3. *Choice of metal electrodes*

Considering the rather high constraining conditions for the choice of these electrodes (as explained above), it appears that there are only a limited number of materials that can be considered as electrode material due to the need of CMOS compatibility. However, the use of an OLED structure with electrically doped charge transport layers extends these possibilities. Otherwise, thin layer such as copper phthalocyanine (CuPc), can be added to facilitate the injection of carriers by presenting a low barrier [LEV 02].

An electrode made up of a stack of aluminum and TiN, as shown in Figure 6.2, fulfills these requirements [KRE 07] and is widely used. This electrode has a roughness of 1.5 nm rms with peaks less than 22 nm. The step height between the electrodes and the isolating silicon dioxide varied from 75 to 90 nm [KRE 07].

On the one hand, the thickness of the TiN layer must be thin to get a high reflectivity of the bottom electrode in the visible range of light. On the other hand, it has to be thick enough to prevent oxidation of the aluminium electrode material.

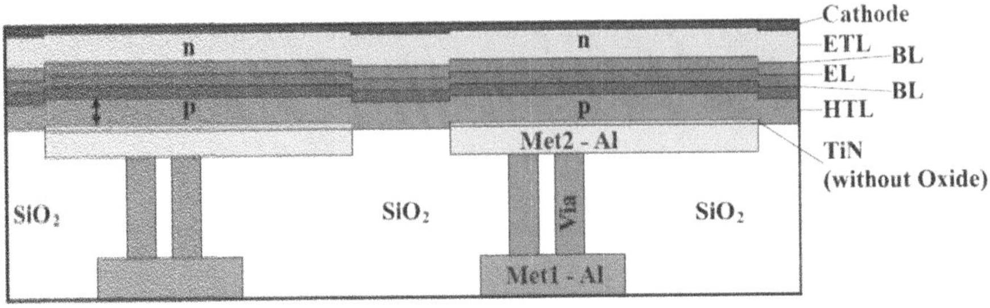

Figure 6.2. *Schematic cross-section of upper part of CMOS active matrix for OLED microdisplay, showing Al + TiN pixel electrodes covered by subsequent OLED stack [KRE 07]. (Reprinted courtesy of SID)*

6.1.4. *Pixel pitch and fill factor*

As stated above, the target for high resolution/small pitch and high performance display is to achieve the smallest pixel pitch together with the smallest electrode spacing, namely the highest fill factor (open aperture). Color pixel pitch as small as 8.1 × 8.1 µm for a 0.48-in. diagonal could be obtained in stripe arrangement, which corresponds to 2.7 × 8.1 µm sub-pixel, and a fill factor of 66 % [GOS 12]. With the same approach, a 9.6 × 9.6 µm color pixel pitch (for a 0.48-in. diagonal) gives a fill factor increased to 71% [GOS 12]. For a 0.5-in., XGA (1024×RGB×768 dots) OLED microdisplay, 9.9 × 9.9 µm color pitch was obtained [ONO 12] as seen in Table 6.1.

Micro-OLED	XGA	720p
Panel size	Diagonal 12.7 mm 0.5-inch	Diagonal 18.0 mm 0.7-inch
Resolution	XGA 2.4M dots (1024xRGBx768)	720p 2.8M dots (1280xRGBx720)
Pixel Size	3.3 µm x 9.9 µm	4.0 µm x 12.0 µm

Table 6.1. *Display specification including resolution and pixel pitch for two OLED microdisplay (from [ONO 12]). (Reprinted courtesy of SID)*

Using the quad color RBGB approach, a color pixel pitch of 9.4 × 9.4 µm could be obtained (4.7 × 4.7 µm sub-pixel) on a 0.61" SXGA microdisplay [HAA 12], and very recently, a 0.39" diagonal XGA (1024×768×RGBW) microdisplay with a color pixel pitch of 7.6 × 7.6 µm (3.8 × 3.8 µm sub-pixel) has been reported in [HAA 13].

6.1.5. *Choice of baseline CMOS circuit*

The CMOS circuit is generally supplied by a foundry. The designer of the OLED microdisplay might find a foundry with a catalog of baseline CMOS technology compatible with the display specifications. A wide choice of foundries and technologies is available. As an example, for a high resolution OLED microdisplay (2560 × 2048 pixels, 4 times SXGA), embedding an all-digital interface (10-bits depth), and needing a high complexity of the digital to analog conversion of the digital video, a 0.13 µm CMOS technology is suitable [GOH 11]. This circuit supplied a 3.3 V power supply. For an OLED microdisplay with a resolution of nearly 4 Mpixels (1280(x3) × 1024), a 0.18-µm CMOS backplane is suitable [GOS 12].

6.2. OLED process on CMOS circuit

6.2.1. *Cluster tool and process*

6.2.1.1. *Concept/architecture*

An OLED deposition machine aimed at handling Si wafer basically consists of a central chamber housing a robot. The robot picks up and sets the substrates from and into the chambers distributed around the central chamber.

Most of these chambers are dedicated to OLED and metal deposition. In addition, a load-lock chamber, a plasma treatment chamber and a buffer chamber are connected to the robot chamber (Figure 6.3).

6.2.1.2. *Deposition chamber*

At the top of the chambers, the shadow mask and the substrate are placed on frames independent of each other. The shadow mask is placed first. On top of it, the substrate is introduced face down. Both of them bear keys that should be aligned with each other by moving the substrate with respect to the mask. During this operation, the mask and the substrate are kept at a few tens of micrometers away from each other. Once the keys are aligned, they are put in contact. A magnet is then put down against the rear side of the substrate so as to attract the mask net and ensure a close contact between the patterns of the mask and the one of the substrate.

A shutter placed below the shadow mask protects the substrate from undesired deposition and stops the deposition precisely when the required thickness of organic material is reached.

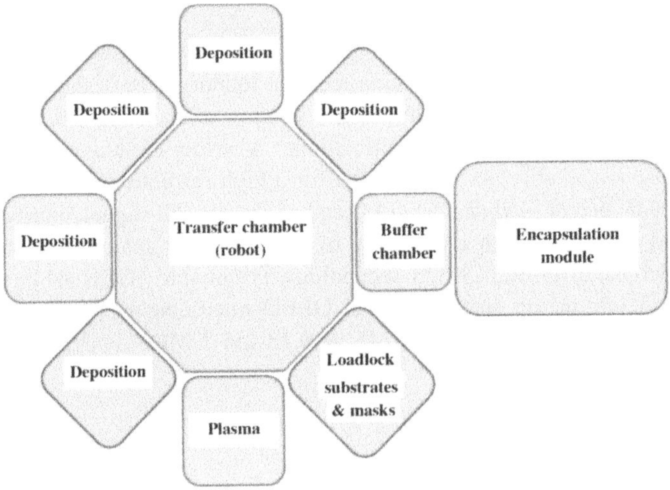

Figure 6.3. *Typical aspect of a cluster tool for OLED deposition onto small substrates*

The main difference between a regular OLED deposition machine and the one made for microdisplay handling is the alignment system for the shadow masks. The system usually consists of CCD camera located on top of the chambers, looking inside through a window to monitor the alignment of the align keys of the glass substrate with the one of the shadow mask. With respect to the camera, the mask is located below the substrate. Therefore, with an opaque Si substrate a periscopic set-up is necessary to visualize the align keys of the substrate through the mask from the bottom. The set-up should be withdrawn before the deposition process in order to avoid any shadowing effect that would impede the uniformity of the deposited film.

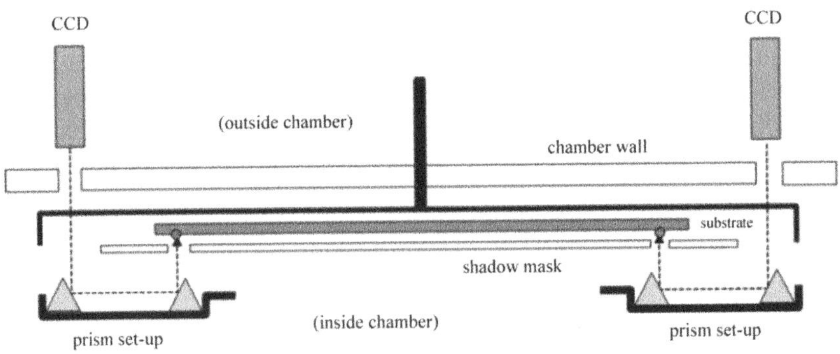

Figure 6.4. *Principle of alignment set-up for an opaque substrate. Two prisms on each side reflect the image of the align keys of the mask and of the substrate toward the CCD cameras. The prisms are withdrawn before the deposition process*

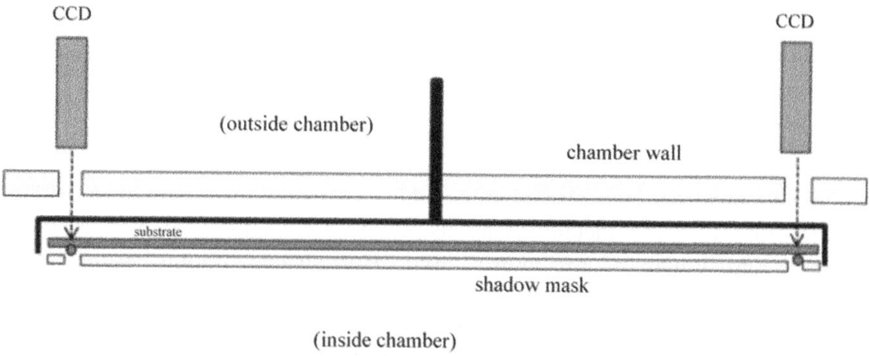

Figure 6.5. *Principle of alignment set-up for a transparent substrate. The CCD cameras look directly through the substrate to align the keys of the substrate and of the mask*

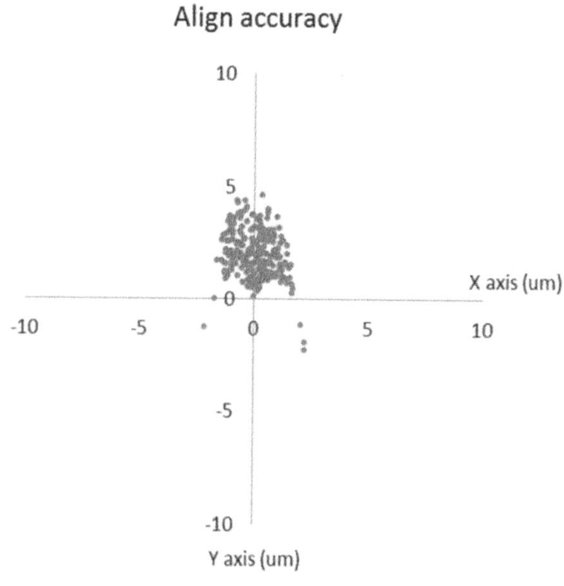

Figure 6.6. *Repeatability of alignment of a shadow mask with a substrate. Accuracy of less than 5 µm can be achieved. (Courtesy of Sunic System)*

6.2.2. *Evaporation sources*

The crucibles filled with the organic materials (powder) are located at the bottom of the chamber. For a better production efficiency, only one material (or a host and its dopant) is usually placed in each chamber. However, according to the duration of deposition of each layer, several materials can be deposited in the same chamber. This is usually the case in R&D equipment.

6.2.2.1. *Crucibles and effusion cells*

Organic materials are placed in crucibles currently made up of Al2O3 or pyrolitic graphite. As for metallic materials (Al, Ag, Ca for instance), pyrolitic boron nitride (PBN) crucibles are mostly used. Boat-shaped vessels made up of molybdenum or tantalum can also be used for metallic materials or for materials with a high temperature of evaporation. However, these latter ones are usually of small content. Moreover, metals, like aluminum for instance, tend to wet the walls of the vessels and spill out. Crucibles are then preferable. Crucibles are of a content of a few cm^3 up to several hundred cm^3.

Crucibles are set in effusion cells aimed at heating the material. These cells consist of a heating set-up, a shutter, a cooling shroud, a thermocouple and a shutter.

The heating is currently provided by a net of heating wires surrounding the crucible. Ceramic rings hold the wires regularly spaced out and keep the crucible away from them.

The cooling system, the thermocouple and the heating netform the set-up that control the thermal cycle. With relevant parameters, the proportional-integral-derivative (PID) control system can regulate the temperature with a typical accuracy of +/–0.02 K

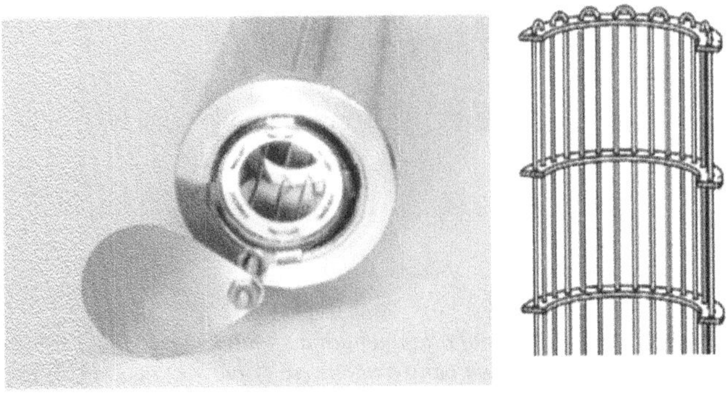

Figure 6.7. *Top view of an effusion cell (shutter open) (left) and heating net (right)*

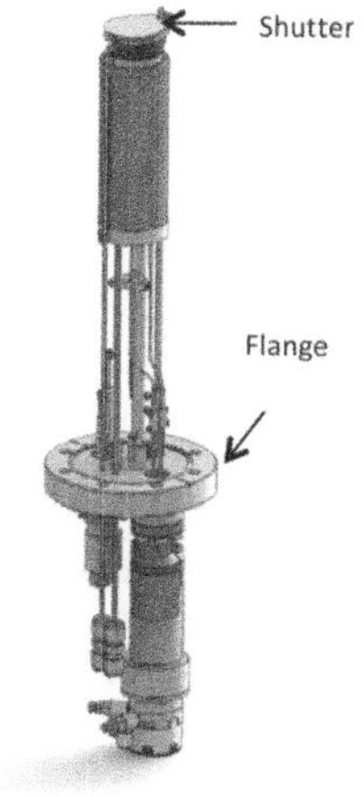

Figure 6.8. *The effusion cells are installed at the bottom of the deposition chambers by mean of a flange*

The shutter is currently integrated to the effusion cell as well as the water cooling system (www.riber.com).

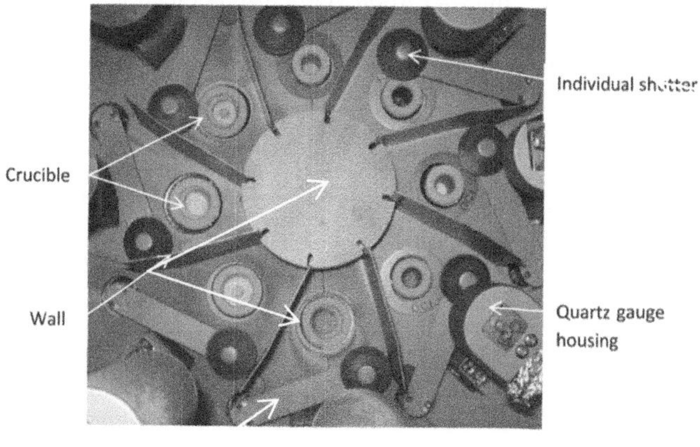

Figure 6.9. *Top view of the bottom of an organic deposition chamber of an R&D cluster tool. Crucibles are separated from their neighboring ones by a wall, as such preventing cross contamination. They are located off axis. Rotation of the substrate is then mandatory to ensure a good uniformity of the deposition*

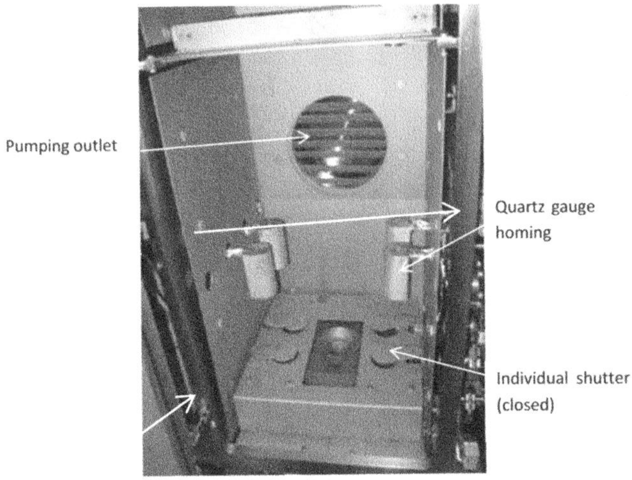

Figure 6.10. *View of metal deposition chamber*

Materials with a high temperature of evaporation are put in so-called high-temperature sources. The vessel itself is heated up by the Joules effect as a part of the power supply circuit. In order to prevent clusters ejection from the source, the vessel can be designed, as such exhibiting a double-bend configuration. The vessel is directly connected to the power supply circuit.

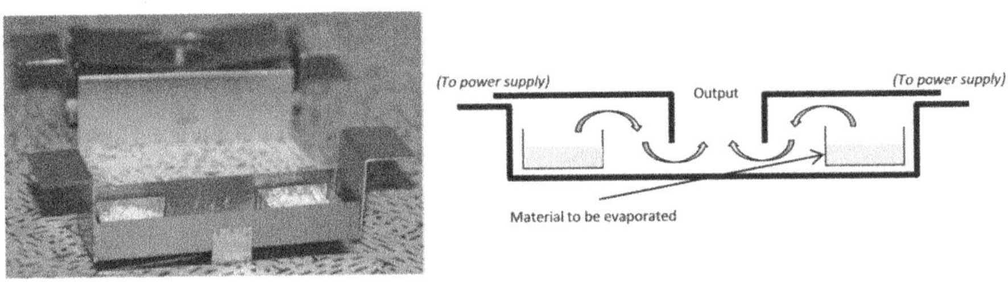

Figure 6.11. *Example of high-temperature evaporation cell*

6.2.2.2. *Ultimate vacuum*

Since the OLEDs are sensitive to oxygen and moisture, the deposition process should be performed in a high vacuum environment: a residual pressure of about 10^{-6} Torr at least is required.

Moreover, such a low pressure prevents parasitic convection of the ambient likely to disturb the "cloud" of organic material being deposited.

Each chamber is then equipped with a primary pump (usually a rotary dry pump) and secondary one (usually a cryogenic pump or a turbo-molecular one).

A parameter to be watched is the rate of pressure rise. In case of temporary failure of the pumping system, a low rate of pressure rise would prevent the substrate in progress from being corrupted.

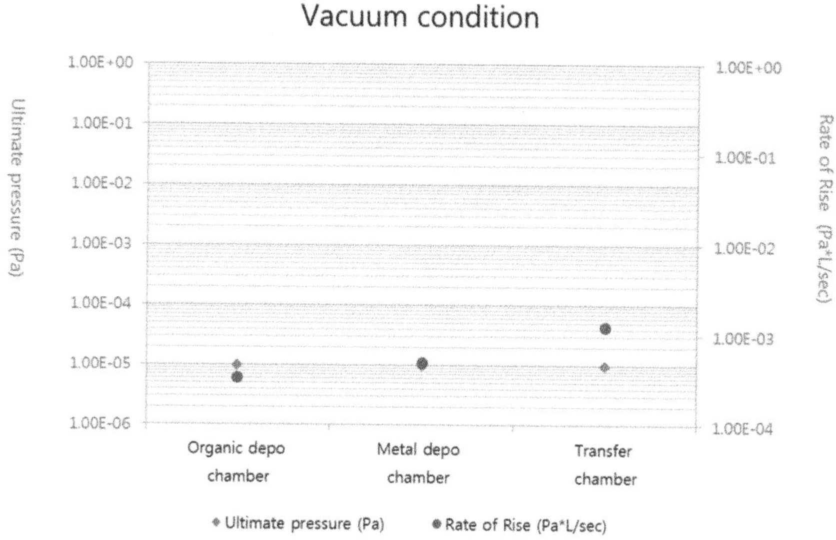

Figure 6.12. *Typical pressure and rate of rise in the chambers of a cluster tool. (Courtesy of Sunic System)*

6.2.3. *Load-lock chamber*

Masks and substrates are first introduced in the load-lock chamber. They are placed in a mobile rack moving up and down in order to align the mask or substrate to be picked up with the arm of the robot. Once placed in the rack, the chamber is closed and the vacuum is set at the same value as that of the central chamber.

The robot picks up the masks and installs them, one after another, in the relevant chamber. Then, each substrate is placed in the plasma chamber in order to undergo a cleaning treatment.

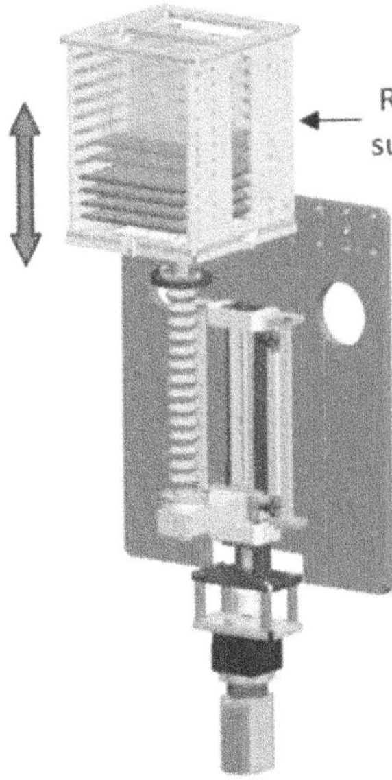

Figure 6.13. *Principle of moving rack in the load-lock chamber. (Courtesy of Sunic System)*

6.2.4. *Plasma treatment*

Prior to the OLED deposition, O_2 plasma cleaning should be performed. This treatment is aimed at removing organic pollution from the surface of the bottom electrodes of the future diodes. A precise control is then critical. Without such a cleaning, the light emission of the OLED structure would at least appear not uniform and at most would not occur.

6.2.5. *Deposition process*

Thermal cycle

Mastering the heating of the organic material in the crucible is the key. Each step of the heating phase should be carefully considered. The heating process starts right with the heating of the crucible from room temperature. Heating the material for several hours under vacuum helps the desorption of humidity out of the powder. Then, the temperature is slowly increased up to a so-called stand-by temperature at which the evaporation is on the verge of occurring, typically about 90% of the evaporation temperature. This step helps to stabilize the deposition rate in a short time once the deposition starts. Indeed, some materials could evaporate more or less violently according to the temperature rise rate due to the grain size of the powder as well as to the nature of the material: some sublimate others evaporate from the liquid phase. Sudden evaporation leads to the deposition of particles detrimental to the uniformity of the film.

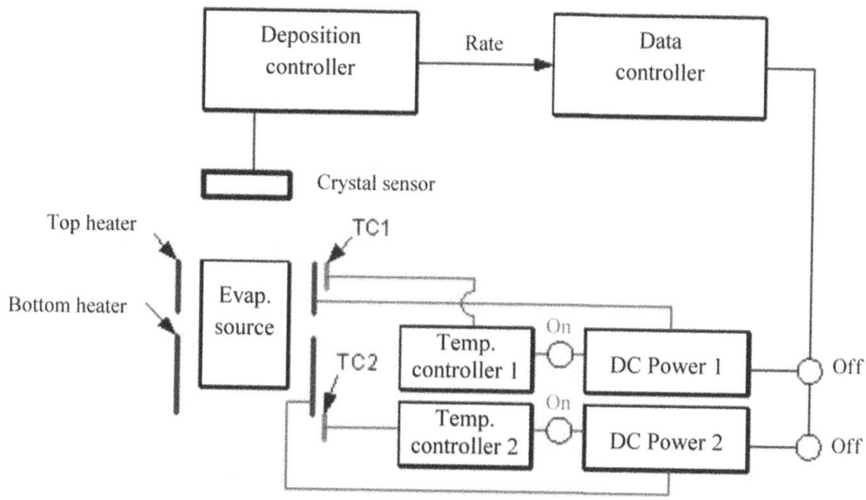

Figure 6.14. *Schematic of the feed-back loop operating during the temperature rise phase. Only the thermocouples are involved in the regulation. (Courtesy of Sunic System)*

This first phase is controlled by the thermocouples and the temperature controller acting on the power supply in order to regulate the temperature.

In order to better regulate the deposition rate, it is then critical to trigger the evaporation smoothly by slowly increasing the temperature from the stand-by point up to the evaporation one. Moreover, such a way of considerably reducing the process time, since the temperature increase to reach the evaporation point can be reached in few seconds. Each crucible is equipped with an individual shutter, and a main shutter protects the substrate once placed in the deposition chamber. These shutters are used in correlation with the thermal cycles in order to ensure a precise control of the deposition rate and the thickness of the layers. During the ramping phase and as long as the crucible is kept at a stand-by temperature or less, all shutters are closed. Then, the deposition phase is two-fold: as the temperature rises up to the evaporation point, the individual shutter is open. The quartz gauge placed at the edge of the crucible, monitors the evaporation. When the evaporation reaches about 90% of the target rate or when the temperature reaches a given value, the PID control takes over. The PID ensures a control loop feed-back mechanism. The controller attempts to minimize the error between the value measured and the set one, by adjusting the power supply of the heaters. The PID leads the evaporation until the target value, the main shutter is opened and the deposition starts. During the deposition, the PID controls the evaporation rates as well as the values of the maximum and minimum power set such as preventing over-(and 'under') shoots.

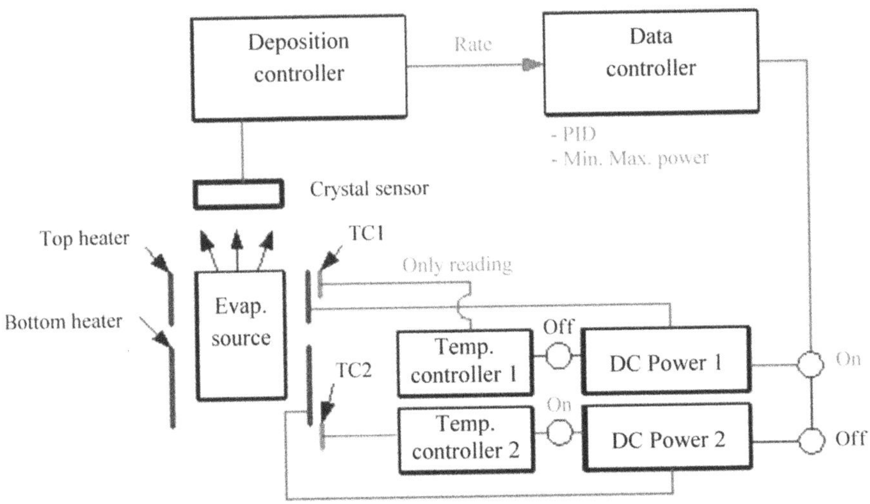

Figure 6.15. *During the deposition phase, the rate is controlled by the quartz gauges. The thermocouples only record the temperature. The PID parameters are set in the controller that acts on the power supplies of the heaters. (Courtesy of Sunic System)*

At the end of the deposition of a given layer, the temperature can be reduced down to the stand-by value until a new substrate is put in place. However, in production, instead of reducing the temperature, the evaporation is kept going on, but the shutters are then closed during the substrate shift.

The heating of the crucibles is ensured by resistive coils into which the crucibles are inserted. One or two coils can be used. The two coils, placed around the top and the bottom of the crucible, can be operated independently in order to ensure a better control of the distribution of the heat, especially as the level of material in the crucible varies. Indeed, bad control of the heating of the crucible can lead to plugging it.

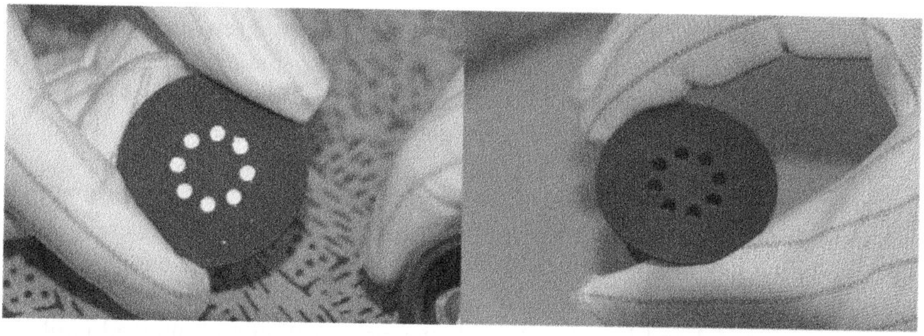

Figure 6.16. *Crucible output plugged by organic material due to a bad control of the heating over the whole crucible. The picture on the right shows the cap before use*

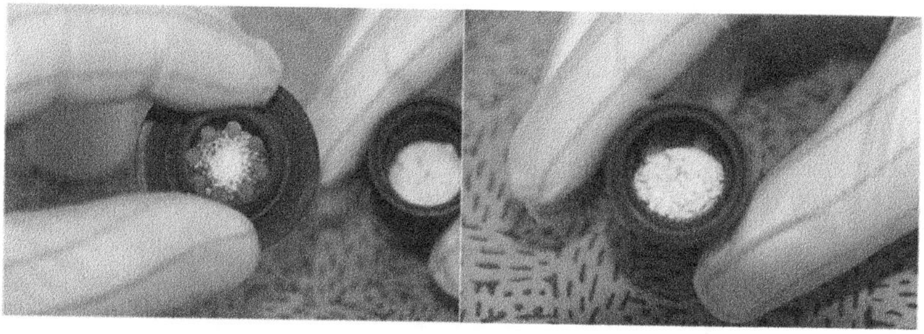

Figure 6.17. *Organic materials can condensate at the output of the crucible and fall back in the crucible in forms of pellets which can lead to cluster deposition onto the substrate. The shape of the crucible along with heating set-up used may not be adapted to this material*

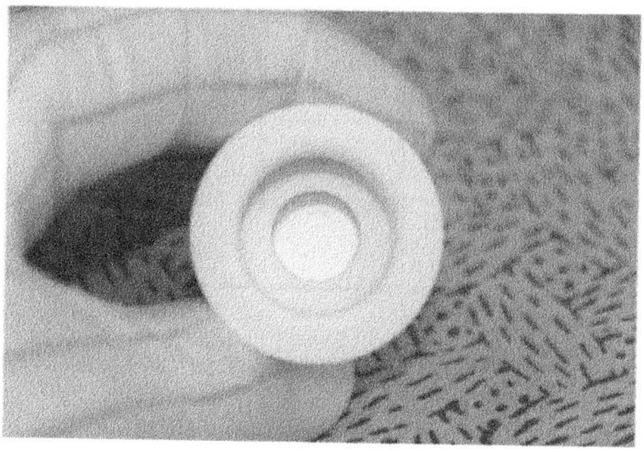

Figure 6.18. *In another type of crucible, the evaporation turns out to keep the powder in a regular aspect*

These crucibles remain open when the shutter is closed a few millimeters above them. The material deposited onto the bottom side of the shutter is also likely to fall back into the crucible in forms of clusters. Moreover, refilling this kind of crucible implies opening the chamber and then breaking the vacuum and the thermal cycle.

Another design of the whole set-up, chamber, crucible and heater turns out to be much more convenient. The bottom wall of the deposition chamber is crossed by a duct closed by a valve. The valve is located in the chamber and is equipped with a heater. The crucible is advantageously replaced by an airtight bottle that is filled with organic material and inert gas for a better preservation and plugged until it is used. The neck of the bottle is compatible with the duct and then enables a connection without breaking the vacuum.

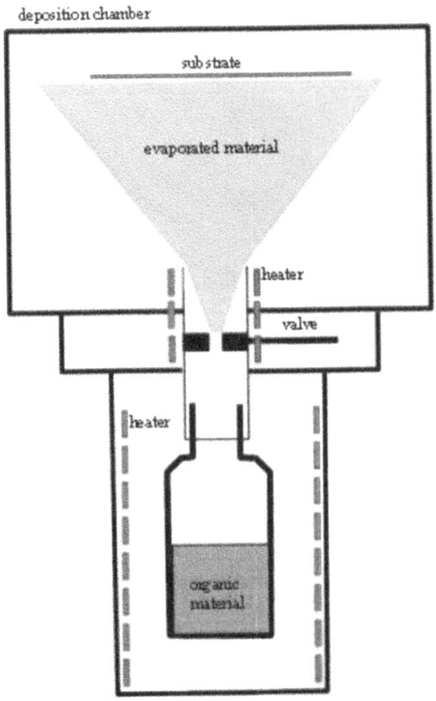

Figure 6.19. *Principle of recharging reservoir (after Guyaux et al. [GUY XX])*

6.2.6. *Thickness, uniformity control and spitting*

During the deposition, the thickness is continuously controlled using a piezo-electric quartz sensor. Located above each crucible but off-axis, the surface of the sensor receives part of the evaporated material. This deposition induces a shift in the frequency of the oscillations imposed to the quartz. This shift is related to the thickness deposited by means of a calibration.

The sensor is the key element of a feed-back loop that regulates the deposition rate. It ensures a real-time monitoring of the deposition, which is mandatory, since the quality of the deposited layer is closely related to the deposition rate. Indeed, according to the material, sudden changes in the evaporation rate or 'high' deposition rate could reflect an ejection of clusters of material: particles are ejected from the crucible and projected onto the substrate that will appear as dark spots on the display. The deposition is currently performed at a rate of 1–10 A/s according to the set-up.

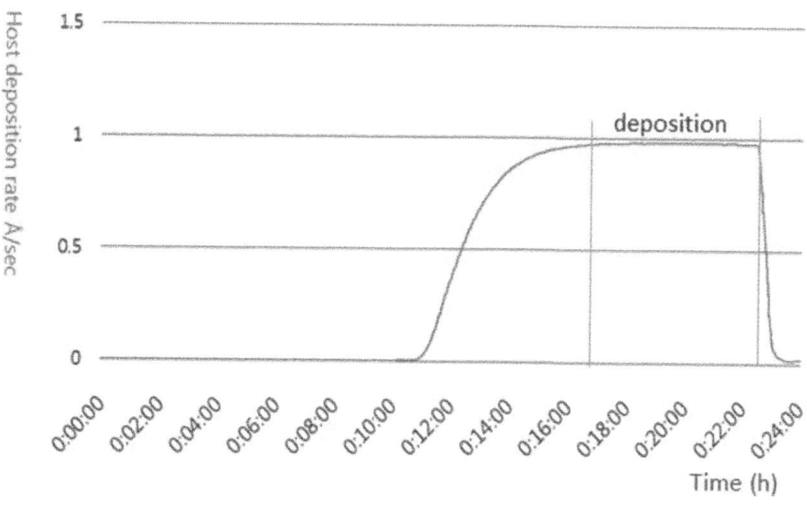

Figure 6.20. *Typical rise in the deposition rate of a host material recorded by the individual quartz sensor. (Courtesy of Sunic System)*

The figure above shows a typical rise in the deposition rate of a host material recorded by the individual quartz sensor. The desired rate of 1 A/s is gradually achieved and kept stable. The deposition onto the substrate occurs from 17 to 23 min. The smooth aspect of the figure is due to the filtering of the data pictured. Such way of displaying this kind of data deprives of numbers of information.

A second sensor is located at the edge of the substrate. This one monitors the material deposited onto the substrate and stops the deposition by triggering the closing of the main shutter when the desired thickness is reached.

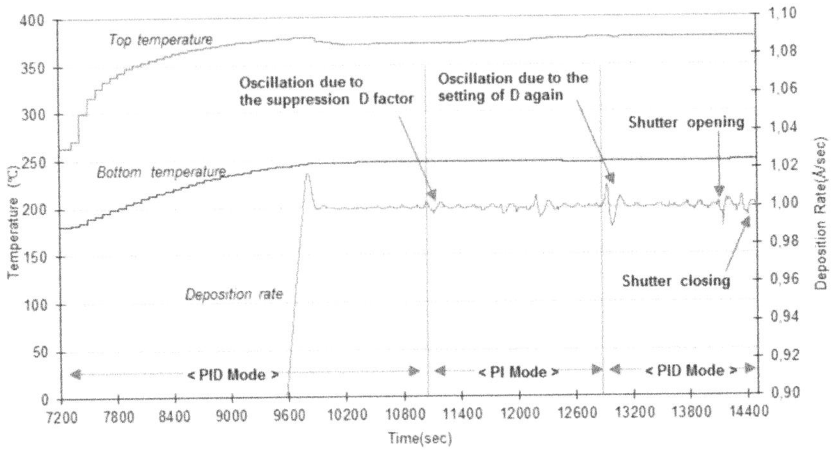

Figure 6.21. *Typical figure of raw data of top and bottom temperature along with the deposition rate versus time. This figure points out the artifacts induced by various operations: change in the PID setting, opening and closing of the main shutter. This gives a clear view of the overshoot and of the stability of the deposition rate. Displaying raw data is usually preferred against the filtered one*

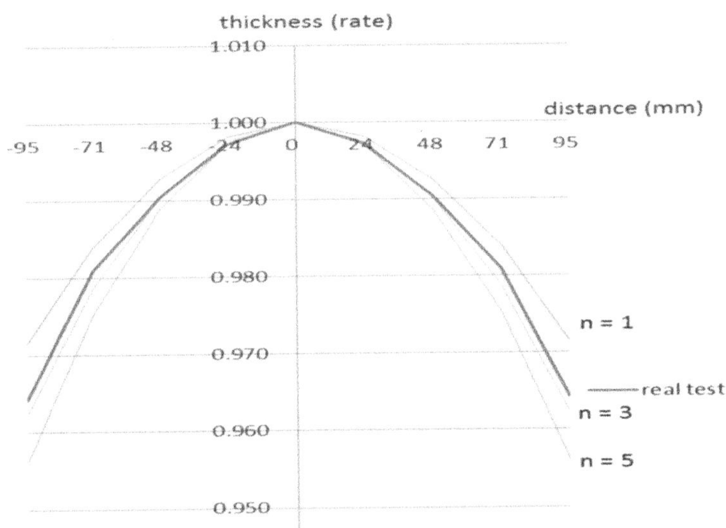

Figure 6.22. *Record of the relative thickness of a film measured on the 200 mm Si substrate compared with the simulated one. The uniformity measured is of +/– 2% (red line) as it is calculated at +/–2.14% (figure n = 3) over an area of 190 mm of diameter. 'n' stands for the power of a cosine function in the equation used to describe the distribution the material in the cloud. (Courtesy of Sunic System)*

Deposition rate, uniformity and quality of the film are closely related to each other. Indeed, the shape of the vapor "plume" depends on the rate of ejection of the material from the crucible. (Shape and tilt of the crucible also play a role in this shape). High evaporation rate leads a directive plume, as a wide one is recommended in order to cover the substrate. Since the substrate is usually rotating (at about 10 rpm) during the deposition some non-uniformities can be reduced, but a narrow plume (due to high evaporation rate) leads to a ring-shaped thickness distribution or to a higher thickness at the center of the substrate. The dynamic of the vapor cloud is nowadays well–known, and the resulting thickness of the deposited film can be simulated with a good accuracy [LEE 09].

As shown in the figure below, the tilt of the crucible also plays a great role in the distribution of the thickness.

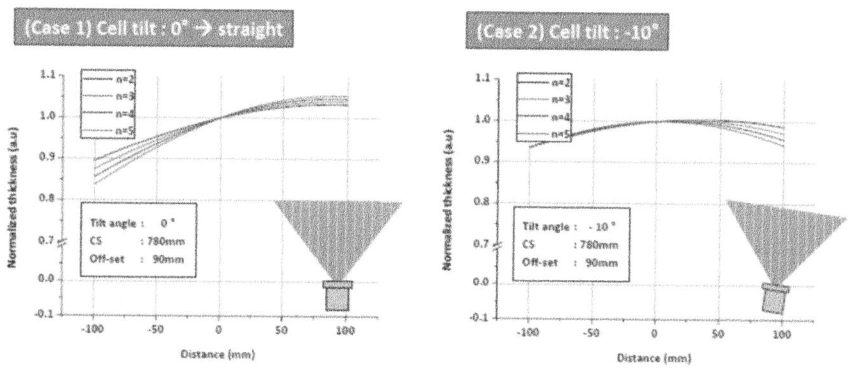

n-factor	Thickness uniformity	
	-100mm ~ 100mm	-75mm ~ 75 mm
2	7.02 %	5.35 %
3	8.39 %	6.40 %
4	9.74 %	7.45 %
5	11.07 %	8.48 %

n-factor	Thickness uniformity	
	-100mm ~ 100mm	-75mm ~ 75 mm
2	3.55 %	2.40 %
3	3.44 %	2.20 %
4	3.38 %	2.07 %
5	3.37 %	1.98 %

Figure 6.23. *Theoretical calculation of the thickness distribution onto a 200 mm substrate for a crucible in a straight position (0° tilt) and for a tilted one. Tilting the crucible improves then the uniformity but at the expenses of the deposition efficiency. (Courtesy of Sunic System)*

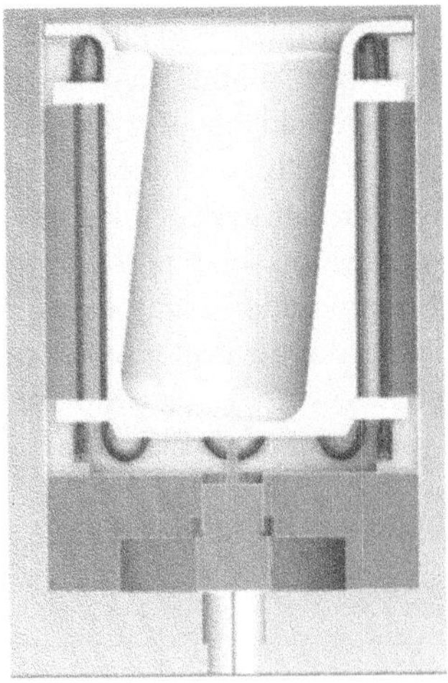

Figure 6.24. *Crucible can be designed such as having a cavity tilted with respect to the main axis, in order to improve the uniformity of the deposition*

The location of the crucible with respect to the substrate, or offset, is also a parameter to take into account in the resulting distribution of thickness.

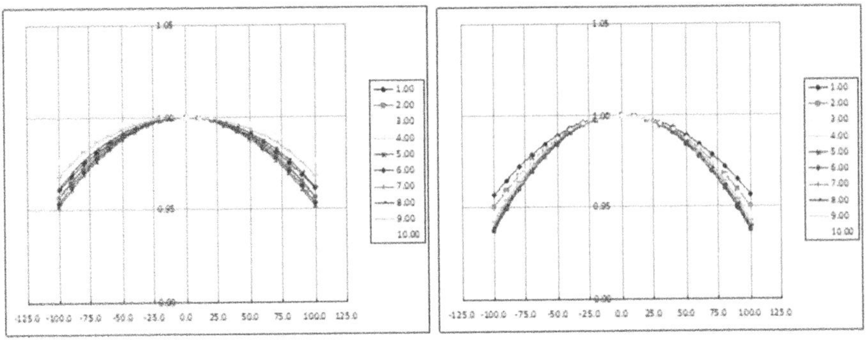

Figure 6.25. *Calculated thickness distribution over a 200 mm substrate for two values of crucible offset (distance of the crucible from the vertical of the center of the substrate) (Courtesy of Sunic System)*

In addition to the uniformity itself, the spitting, i.e. the deposition of particles or cluster of material, is another issue to be considered. As previously quoted, particle ejections can occur in many different cases. The figure hereafter shows not only the influence of the deposition rate but also of the nature and shape of the crucible on the resulting density of particles on the substrate for different organic materials.

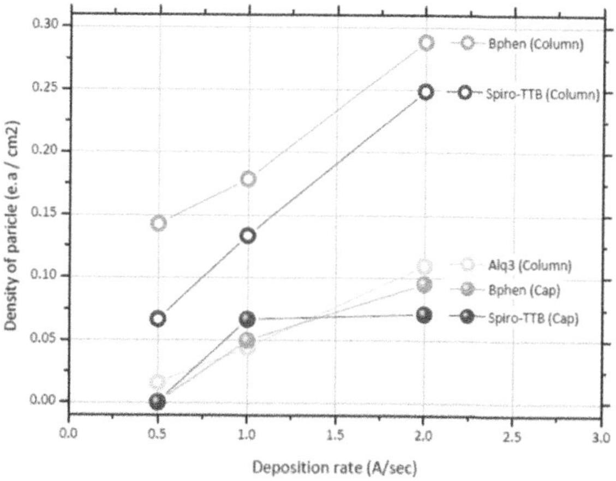

Figure 6.26. *For a given deposition rate and material, the density of particles noticed can vary by about a factor 5 according to the crucible used. (Courtesy of Sunic System)*

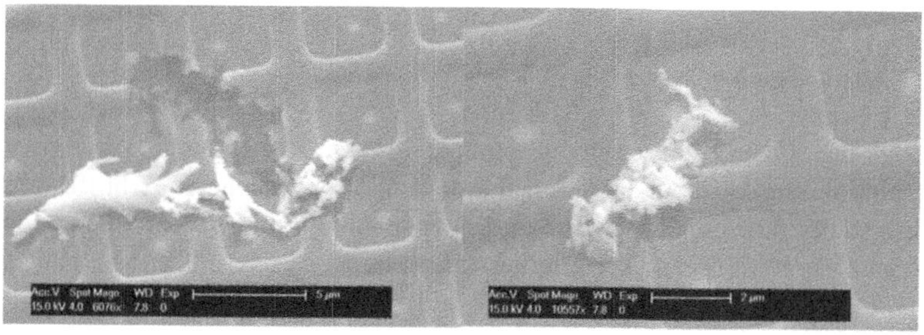

Figure 6.27. *Aspect and shadowing effect of particles present at the surface of the substrate before organic layer deposition or ejected from the crucible*

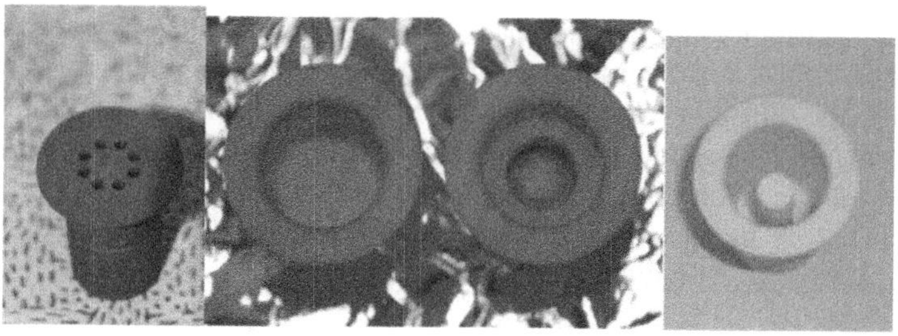

Figure 6.28. *Capped, thin and thick wall or 'column' type of crucibles: the column in the latter one helps to conduct the heat in the bulk of the material ensuring a more uniform heating. Crucible should be adapted to the material and the heating set-up*

6.2.7. *Shadow mask*

A shadow mask consists of a sheet of nickel of about 20 µm etched by electro-erosion in order to open holes of the size of the sub-pixel. This so-called stencil, is stretched and welded or glued on a rigid frame to ensure its flatness and ease of handling. This thin sheet is reinforced by thicker metal around the deposition area. However, according to the size of the sub-pixel with respect to the thickness of the metal sheet, the deposition of individual pixels by this mean is limited to a sub-pixel size of about 40 µm. For smaller pixels, the shadowing effect induced by the metal sheet thickness would prevent a uniform deposition over the sub-pixel area. Therefore, for microdisplays, the holes in the shadow mask are of the size of the whole active area of the display itself, as the deposition of a white emitting structure is mandatory (see "color management" section). This in fact releases constrains on the mask fabrication and alignment accuracy. Indeed, the mask can be made in a 100 µm thick sheet of metal (INVAR for instance, which exhibits a very good thermal stability) and then be much less fragile than a mask made for sub-pixel deposition.

6.2.8. *Buffer chamber*

Once the OLED structure is completed, the substrate is driven to the encapsulation step without breaking the controlled atmosphere. A buffer chamber ensures usually the connection with the encapsulation module where the atmosphere is kept at 0.1 ppm of O_2 and 0.1 ppm of water vapor.

6.3. Encapsulation process

6.3.1. *Encapsulation tools for production*

OLED-based microdisplays are made onto CMOS-built silicon wafers in cluster tools (as described in the former section). At the end of the process, after cathode processing, the thin film encapsulation must be deposited. As described previously (Chapter 2), ALD technology is well-adapted for this purpose. Mass-production dedicated tools can be directly interfaced to the OLED deposition tool to avoid vacuum break. A nice example of such a tool was recently given by the German company Aixtron (see Figure 6.29).

This system is dedicated to Gen 1 (200×200 mm^2) mother substrate, but we might rationally think that it can be easily retrofitted for a silicon 200 mm diameter substrate. Organic chambers are made up of an OVPD system, which in turn can avoid the spitting of particles. As described, this spitting of particles can be detrimental to device quality, because it can cause localized defects that cannot be released afterward. As a result, even if a highly reliable encapsulation is used to protect the OLED, the final product will show some defects that are of course not acceptable at the microdisplay level of definition. The Aixtron system also embeds an ALD thin-film encapsulation system.

Such dedicated systems, targeted for production, are naturally more expensive.

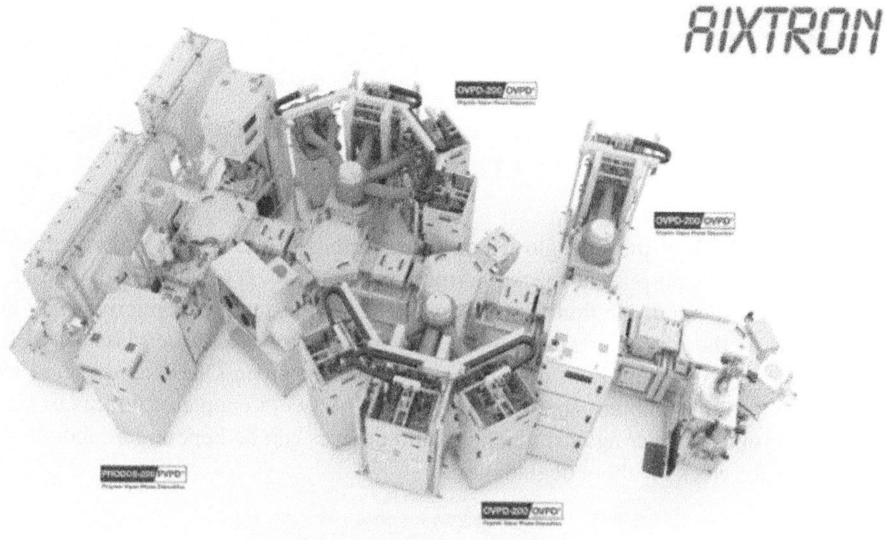

Figure 6.29. *Aixtron systems released in 2014. (Courtesy of Aixtron)*

6.3.2. *Encapsulation tools for pilot line/ R&D*

In an R&D environment, where cost is a major concern and wafer through-put not so important, glove boxes can be used in order to transfer the wafer containing the freshly made, non-encapsulated OLED to the ALD deposition system. A picture of an ALD system (Savannah system from Ultratech/Cambridge Nanotech) embedded into a glove box has been depicted in the following figure, as well as the route to make the encapsulation process.

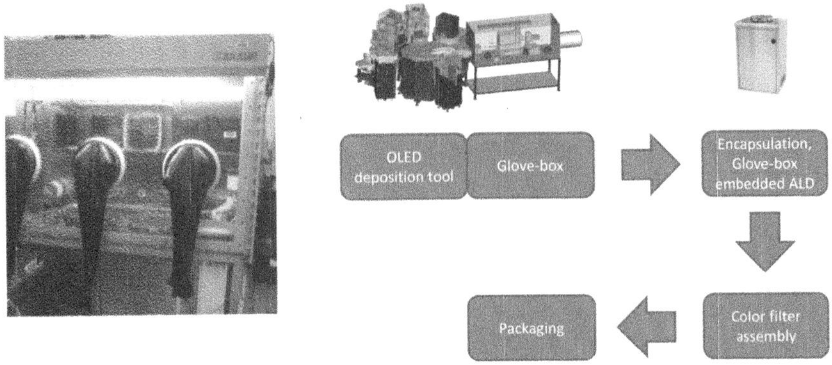

Figure 6.30. *Glove box embedded Savannah ALD system (from Ultratech/Cambridge Nanotech)(left); proposed encapsulation route (R&D environment) (right)*

The transfer from the first glove box to the one with the ALD system is made manually. For this purpose, a special hermetic transfer box is used which is particle controlled. At the end, the thin-film encapsulated OLED sample or microdisplay is transferred to the characterization laboratory in order to make the classical JVL measurements. An example of an Al_2O_3-encapsulated green OLED is depicted in the Figure 6.31. At ambient atmosphere, a single Al_2O_3 (a couple of dozens of nm thick) is a robust barrier against moisture, and devices can be kept under laboratory conditions for weeks without degradation.

Figure 6.31. *High luminance green OLED made onto a silicon wafer. The OLED has been encapsulated with a single 25 nm thick Al_2O_3 oxide layer deposited by low temperature ALD*

For microdisplay products, the top encapsulation must sustain the subsequent technological steps like the color filter assembly. As a result, more robust encapsulation processes could be implemented in the fabrication of microdisplays [MAI 12, SIM 13].

6.4. Color: different approaches and associated processes

6.4.1. *Color management*

According to the size of the sub-pixel, the individual deposition of the OLED RGB structure is not possible due the fact that the size of the pixel is rather small compared with the thickness of the shadow mask. Indeed, a shadow mask usually consists of a sheet of metal of about 20 µm etched by electro-erosion in order to open rectangular holes of the size of the sub-pixel. Such a thickness is obviously not compatible with sub-pixel of about 5 µm of typical size, since the edges of the holes would prevent a correct deposition onto the whole area of the sub-pixels.

The solution currently applied is thus the deposition of a white emitting OLED structure and the assembly of color filters. The white emission should be as broad as possible and should exhibit separated color peaks (orange and blue or red, green and blue).

It has to be noticed that the combination of white OLED and color filters provides a great asset in terms of contrast. Indeed, the ambient light that would reflect on the anode is on the one hand attenuated by the color filter, and on the other hand out of phase with the part of the light reflected by the cathode. This is then an efficient way to increase the contrast of the display without the use of a polarizer, as usually done.

6.4.2. Color filter assembly

Two approaches are to be considered. Color filters can be made on glass and assembled and color filters deposited and etched directly onto the OLED structure.

6.4.2.1. *Color filters on glass*

The main advantage of this approach is that the technology of color filters on glass is already widely spread and used for the mobile phone camera for instance. However, the assembly of pieces of glass bearing the filters onto an OLED is not straightforward.

In order to cover each display of a whole Si wafer with color filters on glass, two ways are to be considered. The wafer scale process or the pick-and-place one.

6.4.2.2. *Wafer scale process*

This process consists of covering the Si wafer with a glass wafer (or color filter wafer) previously coated with glue.

In order to prevent any mismatch of the color filter with the displays, the glass chosen must feature a thermal expansion coefficient close to that of Si like the Corning Eagle XG for instance.

The difficulty also stems from the fact that the contacts of the displays must be kept free of any cap or glue. Since they are usually on the same side of the Si substrate as the display to be protected, the assembly of the cap on the wafer becomes tricky. Grooves are then processed on the glass wafer such that they match the location of the contact on the Si wafer. The glue spread over the glass forms a layer of a few micrometers. The thickness of this layer (typically 5 µm) being much less than the depth of the grooves (typically 200 µm), the glue does not dirty the contact of the display substrate, as the glass wafer is assembled.

However, unless the glue is spread over the glass with the relevant thickness, it is rather difficult to press the substrate to reach a lower thickness than the one deposited: according to the viscosity of the glue and the area onto which the pressure should be exerted uniformly are too high. Since the OLED does not withstand any thermal treatment exceeding 80°C, the glue should be UV curable. The glass color filter is kept in place under pressure when the curing process occurs.

Moreover, once both wafers assembled, the displays must be cut apart. That supposes a two-steps cutting process. On the one hand, along the edges of the display free of contacts, both substrates can be cut from one side of the assembly. On the other hand, the edge of the display bearing the contacts, Si wafer and the glass wafer is cut one after the other from each side of the assembly, the cutting lines being staggered with respect to each other.

6.4.2.3. *Pick-and-place process*

This process implies that color filters are individualized. They must then be cut apart from a substrate. The cutting is known to generate particles that could pollute the color filters. A way to protect the colored photoresist during this process is covering it with a sacrificial photoresist that is removed afterward. This layer must then be cured, as such remaining easy to remove without damaging the underlying colored photoresist.

Then, the pieces of color filter should be glued onto the OLED stack. Glue droplets are first dispensed on the substrate, then, the glass color filter is aligned and pressed against the substrate. This supposes that the OLED is already well-protected by a proper encapsulation.

In order to avoid color shift when looking at the display over a wide range of viewing angles (typically +/– 25°), color filter should be close enough to the emitting medium, i.e. the OLED stack. This implies that the color filter should be pressed onto the microdisplay with a certain pressure in order to reduce the thickness of the glue down to about 1 µm. It then becomes tricky to keep a precision of alignment of about 10% of the sub-pixel size when exerting such a pressure and to ensure the parallelism of the glass piece with respect to the substrate. The viscosity thus becomes a key parameter in order to easily reach such a thickness: glues featuring a viscosity of about 10 cp are to be considered for this step of the fabrication.

The consequence of the thin glue layer required and of the force aimed at pressing the glue out, is the non-negligible risk of punching the OLED structure by particles that may remain between the glass piece and the substrate.

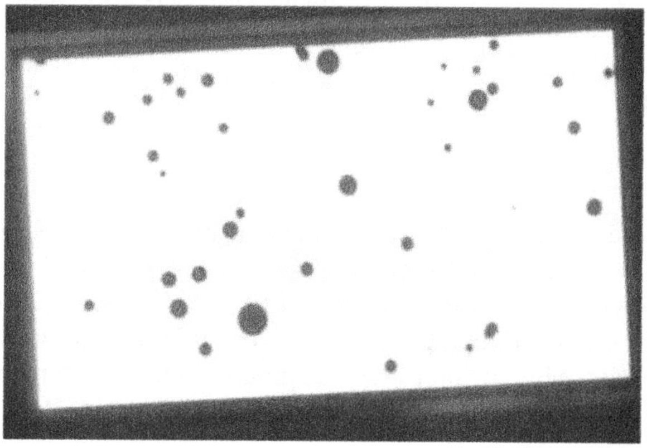

Figure 6.32. *Aspect of a display after color filter assembly onto a substrate bearing particles. The number of dark spots is closely related to the pressure exerted*

Bubbles trapped in the glue are also an issue to be addressed. The glue droplet should be dispensed so as to prevent bubbles to getting trapped when the glue spread over the surface of the display as the pressure is exerted.

6.4.3. *Color filters on OLED*

Another approach consists of making the color filters on the OLED stack, as they are built onto glass, by successive deposition and etching processes for each color. Such a process is well-known on glass, but the issue arising on OLED is the non-complete curing of the photoresist of each color before the process of the following one. Indeed, OLED does not withstand thermal treatment over 80°C, as the photoresist should be cured at 200°C at least. Due to a non-complete curing, the color filters deposited are likely to be damaged by the process of the next one.

However, it is possible to use regular photoresist, but an extra UV curing is necessary to compensate the lack of thermal curing. According to the characteristics of the colored photoresist currently available, the thickness of the filters is of about 1 µm. A two-step deposition (half of the final thickness for each step) is all the more recommended as UV curing is less and less efficient with increasing thickness of photoresist.

So-called "low temperature" color photoresist has also been developed for OLED [PAR 10, KIM 10] or for display on flexible substrates [DEK 06]. Indeed, such displays consisting of the assembly of flexible substrate and flexible color filters would suffer locally from parallax issues when bent. These color filter resists, can be dried at 50°C (instead of 200°C), and the post-bake process can be omitted.

6.5. Packaging

Glass capping

One of the advantages of the color filters on glass is that the glass plays the double role of substrate and of protective cap. Such a protection is indeed highly recommended owing to the fragility of the OLED structure, even if color filters are deposited onto OLED. In the latter case, the constrain of glue thickness is not as severe as for the assembly of color filters on glass onto the structure, but the issue of bubbles trapped in the glue layer remains critical.

As for assembly of color filters, the protective glass cap can be done at a wafer scale with the same constrains stemming from the contacts of the display.

6.6. Display testing and performances

After wafer fabrication, electrical testing is performed. Testing is performed on a dedicated automated probe station. Such equipment is derived from standard semiconductor testing, and adapted for microdisplay testing. In addition to classical electrical characterization, such equipment can evaluate electro-optical performances of the displays, such as luminance versus voltage, luminance uniformity and emission spectra (for more details on OLED characterization, refer to Chapter 4). For this automated testing, OLED microdisplays are placed on a probe-card.

After parametric testing, wafers are singulated using classical dicing equipment. Then, OLED microdisplays can be electrically connected, as will be described in the next section.

Figure 6.33. *TEL P-8XL wafer probe station for testing OLED microdisplays (left) with focus on OLED microdisplay dedicated probe-card (right). (Courtesy of CEA-LETI)*

6.7. Electronics

This section describes how to provide video and configuration data from an image source, e.g. computers and smartphones to microdisplays. The second part of this section shows different technologies to mount and pack microdisplays on modules.

6.7.1. *Data and configuration interface*

To provide images to microdisplays, a video source is needed. Possible video sources are:

– personal computers (PCs);

– notebooks and netbooks;

– tablet PCs;

– personal digital assistants (PDAs);

– mobile computers;

– field programmable gate arrays (FPGAs).

To drive microdisplays from video sources, three different connections are essential: the video data bus, the configuration bus and the power supply connections. The video data bus transfers the images via video data stream to the microdisplay. The process is comparable to the video projector connection on computers. To set internal microdisplay modes, e.g. brightness, pixel mapping and frame rate, a configuration bus is required. Additionally, a power distribution connection is needed to provide operating and core voltages. Some OLED microdisplays need a negative cathode voltage. Table 6.1 shows possible video and configuration interfaces.

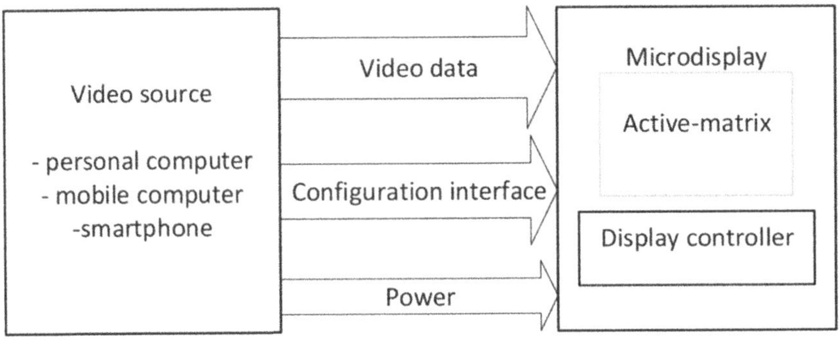

Figure 6.34. *Data and configuration interfaces of microdisplays*

Video interfaces			
Interface		*Data-rate (max.)*	*Note*
VGA	Video graphics array	25 Mbit/s	Analog
DVI	Digital visual interface	6x1.65 Gbit/s	Digital
HDMI	High-definition multimedia interface	10.2 Gbit/s	Digital
MDDI	Mobile display digital interface	3.2 Gbit/s	Digital, for mobile devices
MIPI	Mobile industry processor interface	10 Gbit/s	Digital, for mobile devices
MPL	Mobile pixel link	10 Gbit/s	Digital, for mobile devices
MHL	Mobile high-definition link	10.2 Gbit/s	Digital
DP	Display port	5.4 Gbit/s / lane	Digital
Configuration interfaces			
USB	Universal serial bus	500 MBytes/s	Power supply possible up to 500 mA
I^2C	Inter-integrated circuit	3.4 Mbit/s	
SPI	Serial peripheral interface	600 kbit/s	

Table 6.1. *Video and configuration interfaces*

Currently available microdisplays do not provide an integrated interface controller to support the common video interfaces listed in Table 6.1. Most microdisplays have a parallel digital RGB video interface (raw data) or an analog raw data interface with digital synchronization signals. Figure 6.35 shows a parallel digital RGB video interface between a video source and a microdisplay. On the one hand, there are synchronization signals to signalize the start of new images or rows. On the other hand, each color has its own bus (8–10 bit). The color bus can transmit all data for RGB within one clock cycle in parallel. To transmit a complete image, we need approximately x-resolution times y-resolution pixelclock cycles. The differences to the analog raw data interface are the analog signals for RGB instead of digital busses.

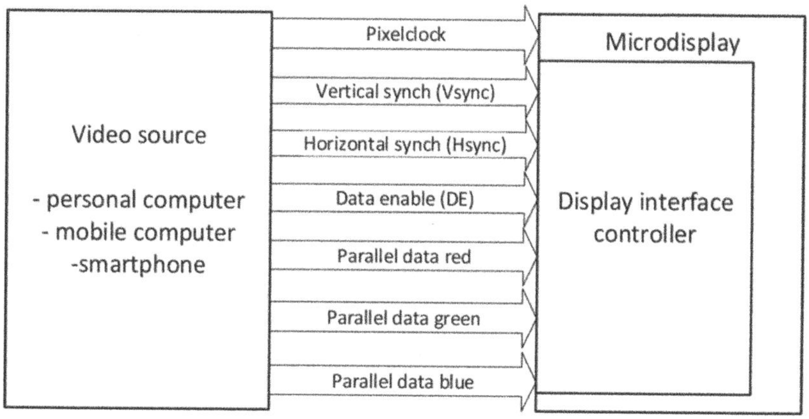

Figure 6.35. *Block diagram of parallel digital RGB video interface*

Figure 6.36 shows the timing scheme of the parallel digital RGB video interface. All signals are synchronized with the rising or falling edge of the pixelclock. The Vsync signal signalizes the start of a new frame. The Hsync signal is the next impulse and signalizes the beginning of a new line within the frame. After the first Hsync signal, there are empty rows (row blanking) before the first row with valid data starts. The first row with valid data also begins with a Hsync signal followed by empty data (column blanking). The rising edge of the data enable (DE) signal indicates the beginning of valid pixel data (3 × [8–10 bit] per pixel clock cycle). The falling edge of the DE signal within the row signalizes the end of valid data followed by additional column blanking. This process is repeated until all rows are transferred. After the last row of valid data, empty rows are also transmitted. After the blanking of these rows, a new frame starts with the next Vsync signal.

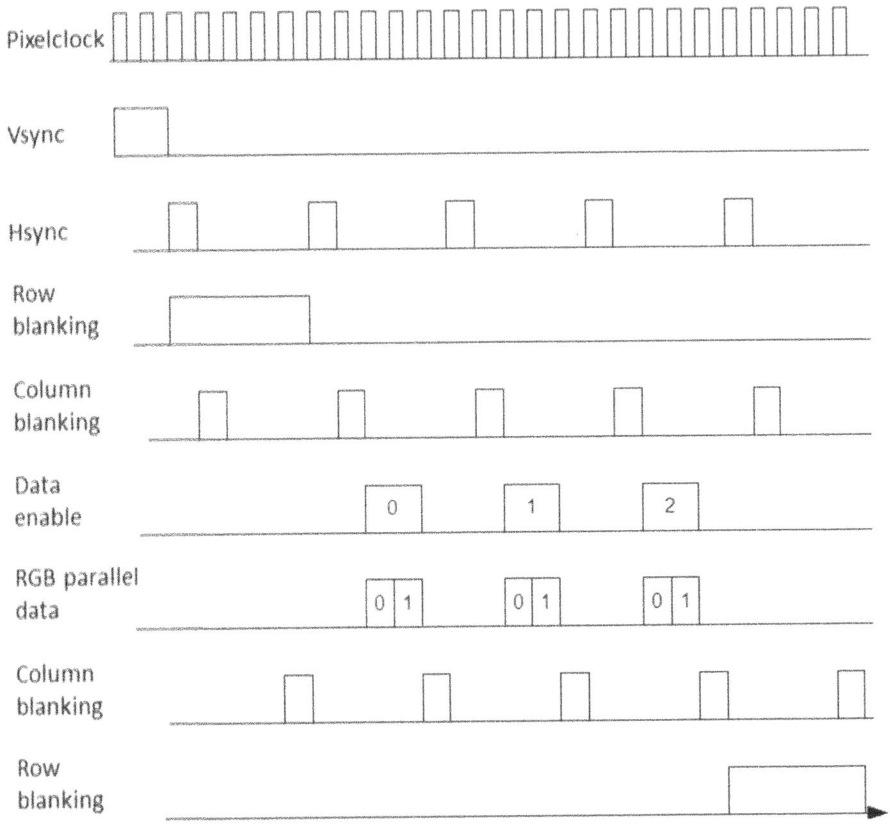

Figure 6.36. *Parallel digital RGB video interface timing scheme*

6.7.2. *Packaging and mounting of microdisplays*

Most microdisplays are provided on CMOS chips. To integrate microdisplays in electronic devices, e.g. data glasses and viewfinders, the microdisplays must be packed on modules. There are three different possible technologies.

1) chip-on-board (COB);

2) chip-on-flex (COF);

3) chip-on-socket (COS).

6.7.2.1. *Chip-on-board (COB)*

The microdisplay is mounted on a printed circuit board (PCB). The PCB includes bondpads. The bondpads are connected to an interface connector and in some cases to active and passive parts. The microdisplay packaging requires four steps:

1) mounting the active and passive parts e.g. capacitors, ICs and connectors;

2) mounting the microdisplay on the board (glueing);

3) bonding the bondwires from the PCB bondpads to the chip bondpads;

4) encapsulation of the bondpads (Glob top encapsulant).

The advantage of this technology is the implementation of the microdisplay and interface ICs on one single board to provide a high packaging density.

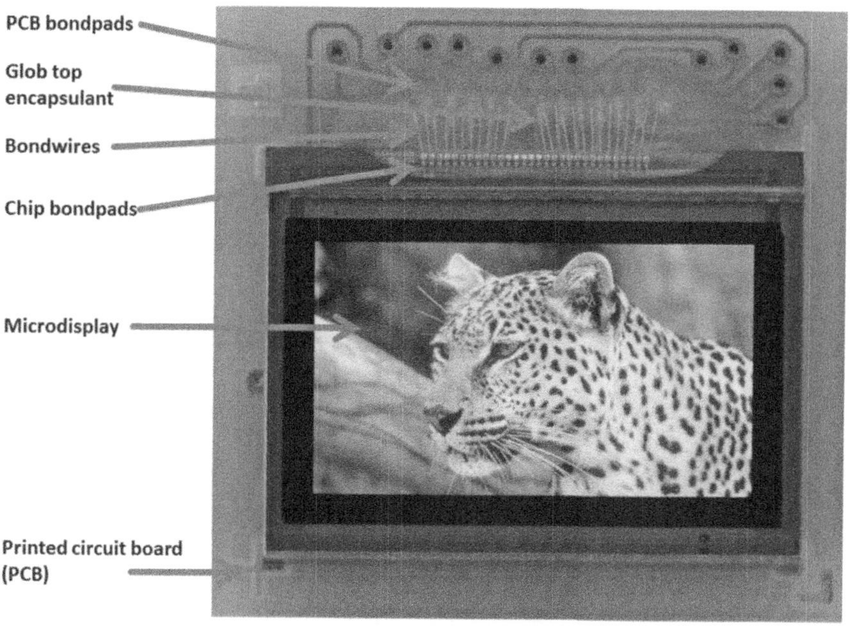

Figure 6.37. *Chip-on-board. (Courtesy of Microoled)*

6.7.2.2. *Chip-on-flex (COF)*

The COF technology is comparable to the COB approach. Within this technology, the microdisplay is mounted and connected on a flex cable. The advantage of this technology is the complex integration in low space devices.

Figure 6.38. *Chip-on-flex*

6.7.2.3. *Chip-on-socket (COS)*

The microdisplay is packed on a standard chip housing e.g. plcc28. By mounting and connecting the microdisplay to the standard package, the steps are similar to the COB technology. Figure 6.39 shows an example of the COS technology. The advantage of this technology is the easy replacement of microdisplays.

Figure 6.39. *Chip-on-socket. (Courtesy of COS)*

6.8. Process and performance evolutions

Performance evolutions of OLED microdisplays will be explained in the final chapter. However, the trend is that according to the targeted applications, power consumption has to be kept as low as possible, and luminance has to be increased (particularly for augmented reality applications in outdoor use). Therefore, efforts should be put on structures and devices that would improve the light extraction from the OLED stack and refine the color purity. This is one of the promising avenues to be explored as shown by the numerous works recently published in this field [SCH 13, FAN 13, LEE 13, HOF 11].

These evolutions concern mainly the OLED device itself, which means materials improvement, modification or addition of layers, etc. Overall, no significant change from the fabrication tools is likely to occur.

Also, at the difference of conventional, medium-to-large size OLED displays, where the trend is to deposit on larger and larger substrates, in the case of microdisplays the substrate size follows the semiconductor industry, which means it is not expected to change soon, and will stick to 200 or 300 mm wafer size. Therefore, the deposition tools will remain similar, only upgrades leading to higher through-put may happen.

6.9. Bibliography

[DEK 06] DE KEYZER G., HOELZLE D., CRACIUN L., *et al.*, "Low temperature colour filter resist for flexible displays", *IDW '06 – Proceedings of the 13th International Display Workshops*, vol. 2, pp. 917–920, 2006.

[FAN 13] FAN Y., CHEN J., MA D., "Enhancement of light extraction of green top-emitting organic light-emitting diodes with refractive index gradually changed coupling layers", *Organic Electronics: Physics, Materials, Applications*, vol. 14, no. 12, pp. 3234–3239, 2013.

[GOH 11] GOHRI V., ESPUNO L., HAAS G., *et al.*, "High-brightness green OLED microdisplay with high-resolution and low power consumption", *Proceedings of Eurodisplay '11, Society For Information Display, Arcachon, France*, pp. 112–15, 2011.

[GOS 12] GOSH A., WACYK I., ALI T., *et al.*, "High-brightness and high-resolution AMOLED microdisplay", *Proceedings of IDW/AD'12*, Japan, 2012.

[GUY 05] GUYAUX J.L., STEMMELEN F., BOUCHAIB P., "Device for vacuum deposition with recharging reservoir and corresponding deposition method", *European Patent*, no. 1, p. 825 018 B1, 2005.

[HAA 12] HAAS G., MARCELLIN-DIBON E., "High resolution OLED microdisplays for electronic vision systems", *Proceedings of IDW/AD'12*, Japan, 2012.

[HAA 13] HAAS G., "High resolution OLED microdisplays – trends in applications and technology", *Plastic Electronics*, Dresden, Germany, 9 October 2013.

[HOF 11] HOFMANN S., THOMSCHKE M., LÜSSEM B., *et al.*, "Top-emitting organic light-emitting diodes", *Optics Express*, vol. 19, no. 23, pp. A1250–A1264, 7 November 2011.

[KIM 10] KIM S., SHIM H., KWON H.-J., *et al.*, "Low temperature color filter on thin film encapsulation active matrix OLED (AMOLED) for flexible display", *IDW''10 – Proceedings of the 17th International Display Workshops*, vol. 2, pp. 1119–1122, 2010.

[KRE 07] KREYE D., TOERKER M., RICHTER B., *et al.*, "Characterization of orange OLED stack for bidirectional active-matrix OLED microdisplays", *SID'07 DIGEST*, pp. 860–63, 2007.

[LEE 09] LEE E., "Simulation of the thin-film thickness distribution for an OLED thermal evaporation process", *Vacuum*, vol. 83, pp. 848–852, 2009.

[LEE 13] LEE J., "Straight-forward control of the degree of micro-cavity effects in organic light-emitting diodes based on a thin striped metal layer", *Organic Electronics: Physics, Materials, Applications*, vol. 14, no. 10, pp. 2444–2450, 2013.

[LEV 02] LEVY G., EVANS W., EBNER J., *et al.*, "An 852 600 pixel OLED-on-silicon color microdisplay using CMOS subthreshold-voltage-scaling current drivers", *IEEE Journal of Solid-State-Circuits*, vol. 37, no. 12, pp. 1979–89, December 2002.

[MAI 12] MAINDRON T., DOYEUX H., KANAAN H., *et al.*, "Lifetime measurement and reliability upon storage of thin-film encapsulated top-emitting PIN OLED", *Proceedings of SID*, 2012.

[ONO 12] ONOYAMA Y., YAMASHITA J., KITAGAWA H., *et al.*, "0.5-inch XGA Micro-OLED display on a silicon backplane with high-definition technologies", *SID 2012 DIGEST*, pp. 950–53, 2012.

[PAR 10] PARK C.Y., LEE S., HYUN C.-H., *et al.*, "High efficient top emitting OLED micro-display for personal display application", *IDW'10 – Proceedings of the 17th International Display Workshops*, vol. 1, pp. 327–330, 2010.

[SCH 13] SCHWAB T., FUCHS C., SCHOLZ R., *et al.*, "Improved light outcoupling and mode analysis of top-emitting OLEDs on periodically corrugated substrates", *Proceedings of SPIE – The International Society for Optical Engineering*, vol. 8829, Article number 88291I, 2013.

[SIM 13] SIMON J.-Y., MAINDRON T., AVENTURIER B., *et al.*, "ALD-based multilayer encapsulation of PIN OLED: on the stability of the organic layer in 85°C/85% RH storage conditions", *ALD2013 Conference Proceedings*, 2013.

7

Applications of OLED Microdisplays

7.1. Introduction

Generally speaking, the main applications for any microdisplay are in the field of near-to-eye or in the miniature projection systems, where the tiny pixelated microdisplay surface is enlarged (or magnified) by a dedicated optics and projected either as a virtual image for near-to-eye application or as a real image on a physical screen for projector applications.

– For projector applications, the collimated optics role is to make an enlarged image of the microdisplay on a physical diffusing screen, so that any user could see the image on a large screen. In these systems, the main parameter is the light flux (or lumen) the system could generate and carry from the microdisplay to the screen. organic light-emitting diodes (OLED) microdisplay is an emissive display. It means that it does not need any external light source to emit the desired light. OLED microdisplay is not intended currently for miniature projector application; since the level of brightness required is out of reach of OLED brightness performances today (requested brightness at microdisplay level is ~x 1,000 what OLED is offering today with 300 Cd/m^2 and 10% projector efficiency delivering 10 lm). Others non-emissive microdisplay technologies such as LCD, LCOS, MEMS-based display such as DLP or scanning Mems technology are used, where the microdisplay is illuminated by an external light sources (Leds or Laser). This is why the main applications of OLED Microdisplay are today geared toward near-to-eye systems.

– For near-to-eye applications the optical system collimates the light from the microdisplay and projects the image near infinity (or far away from the eye). The eye lens then focuses the enlarged image of the microdisplay into the retina. In these systems, the main parameter is the brightness perceived by the human eye. The main

Chapter written by Khaled SARAYEDDINE, Ersun KARTAL and François TEMPLIER.

applications of OLED microdisplays for this type of system are the viewfinder embedded into camera/camcorder or into any industrial or military products and the wearable display devices such as HMD, augmented reality or informative glasses.

In the following sections, we will focus on OLED for near-to-eye applications. Two main parts will address the two main applications of such displays today: head-mounted display systems, and electronic viewfinders. In each case, the specific optical system will be described, together with the contribution of OLED microdisplay. We will describe the main advantages of OLED microdisplay in terms of power consumption, image quality and footprint reduction for the opto-mechanical systems, especially for wearable consumer applications. In each case (HMD and viewfinder products for professional and consumer applications), we will finally stress the technical challenges and evolutions of these systems, to deduce what the desired improved characteristics for future OLED microdisplay are. The last paragraph will list other applications where such displays could be used.

7.2. Head-mounted displays and informative glasses for consumer and professional applications

7.2.1. *General requirements for HMD for consumer and professional markets*

Head mounted display (HMD) has been the focus of military and avionics specialists during the last two decades. The need for HMD integrated into the pilot helmet always in awareness pushed several actors to produce see-through HMD systems for which the virtual image is projected into the helmet visor. These systems are heavy and very expensive. Both monocular and binocular systems were produced.

Recently, several companies and start-ups developed near-to-eye hardware intended for augmented reality and informative glasses for consumer market. Not all these products are optimized in terms of performances and ergonomics and few have learned from the rich military experience over the past decade.

It is difficult to establish full technical requirements for the consumer and professional applications. However, some common features emerged, which we will describe below:

– See-through capability, i.e. full transparency of the HMD for the outdoor scene. This feature is key for HMD use, mainly for mobile consumer use, or for situation awareness in military/avionics or for hands-free professional applications. By see-through, we mean the capability of the display to allow full use of the outdoor scene without any obstruction. Thus, any obstructed display or see-around display are to be excluded.

– The capability of the system to display information in the line of sight of the user's eye in order to have a true augmented reality feature. Thus, we exclude any attempts to display information only in glance mode (outside the line of sight), which limits the power of HMD as a provider of real-time information that overlays over the line of sight of the user.

– The capacity to display large-size images with decent image quality. This implies having a large field of view (FOV), red, green and blue (RGB) color display, as high brightness and high contrast for an image without geometrical distortion.

– Ergonomic form factor. This is a key factor for end users, since these wearable displays should have a low weight (<100 g), small form factor and easy to wear, so as to have a large eye box to allow users of different interpupillary distance to still see the image without any adjustment.

– Perfect angular position adjustment of right and left virtual images for binocular system to avoid eye strain, so as to manage to reduce right and left eye rivality for monocular display system.

– Low-power consumption and high efficiency to allow HMD systems to be used with a low weight and low-power battery.

These general requirements will be derived with more precise parameter values for consumer as for professional applications in the following sections.

As microdisplay OLED is quite a recent technology, current near-to-eye products mainly use other microdisplay technologies, such as liquid crystal display (LCD) and liquid crystal on silicon (LCOS). Thus the challenge for OLED is to replace LCD and LCOS while bringing lower power consumption and better image quality.

7.2.1.1. *Specific requirements for HMD for the consumer market*

In addition to the general requirements mentioned above, the specific requirements for the consumer market are (in order of priority): scalability of display technology and its cost, weight and form factor, power consumption and image quality.

Display technology should be scalable to allow large-volume production with low cost. As any consumer product, the target street price for the glasses should be close to $200, which excludes any exotic optical technology that add cost and yield issue.

The weight is an important factor for customer adoption; since glasses are a wearable product that should have an ergonomic form factor close to normal glasses, with a target weight close to 50 g.

The power consumption factor is related to the efficiency of microdisplay and optical systems, as for the power consumption of embedded electronics. OLED microdisplay offers the best technology for low-power consumption. The efficiency of optical systems will be described in the following paragraph for each optical technology. The embedded electronics system for wireless glasses uses a small battery integrated to the glasses frame. The reduction of total power consumption implies smaller and lower-weight battery, and also a limited use of others embedded sensors, such as GPS and streaming high-definition content from the camera. The target life time for consumer products is about one day (or 4 h full ON), which yields less than 10 g and 8 cm battery with less than 600 mAH power.

Image quality is a key differentiator as it determines the potential use of the glasses for their final applications. A true see-through (or transparent) color display is a fundamental feature for mobile use of glasses for augmented reality and informative applications. Large FOV is important, since smaller virtual image limits the display use to few signs and text content. The display resolution should be adapted to the FOV. Near-to-eye display could then be characterized by the number of pixel per degree, rather than by the native resolution of the display. The requirement for image quality is to have a FOV larger than 20 deg, with an angular resolution of about 40 pixels per degree that is close to the eye resolution limit.

7.2.1.2. *Specific requirements for HMD for the professional market*

It is very difficult to establish a precise list of requirements for all the professional applications, since they are wide and each request a specific form factor and specific environmental conditions.

The most difficult area is the avionic and military applications for which the near-to-eye system is generally fixed to the pilot's helmet. The display is then a part of the helmet equipment and should comply with strict requirements for any wearable equipment in the cockpit. Several papers describe the requirements in this area, which mainly related to reliability, safety and potential use in situation awareness [MEL 11]. Generally speaking, a true see-through near-to-eye system is required, where any limitation of the natural field of the view of the pilot is not accepted. The display should have a very high brightness (up to 10,000 Cd/m^2) for day operation, as the ability to have very high contrast for night vision.

The other industrial applications that have a lot of interest is AR glasses are in the logistic, warehouse, security, maintenance and specific factory equipments. For this kind of application, life time (power consumption) and ergonomics are important factors. The glasses should have a high-definition embedded camera and other localization sensors to allow the user to have hands free operation during several hours.

In the following sections, we will focus on a detailed description of microdisplay-based optical systems. We start by a general description of the collimating optics and describe several HMD techniques for non-see-through and see-through (transparent) near-to-eye systems.

7.2.2. *Optical system architecture for near-to-eye display*

7.2.2.1. *Basic principle of near-to-eye optical systems*

Figure 7.1 shows the basic principle of a classical near-to-eye optical system. There are two types of near-to-eye optical systems; the pupil forming system and the non-pupil forming system. In the pupil forming system, an intermediate enlarged image is located in the light path, which offers some advantage to filter, correct system performances [ROL 05]. However, the pupil forming optical system is quite cumbersome and is rarely used in current HMD consumer products. Non-pupil forming systems, as depicted in Figure 7.1 have the advantage of a short footprint. In this case, the eye pupil matches the exist pupil of the optical system.

– As we can see from Figure 7.1, the light out of an OLED microdisplay located nearly at the focal plane (F) of a collimator lens is collimated and crosses the real exit pupil located at the eye pupil plane. This collimated lens enlarges the microdisplay image and projects the image at a large distance D (few meters to infinity). The eye itself focuses the light into the Retina to form a large virtual image.

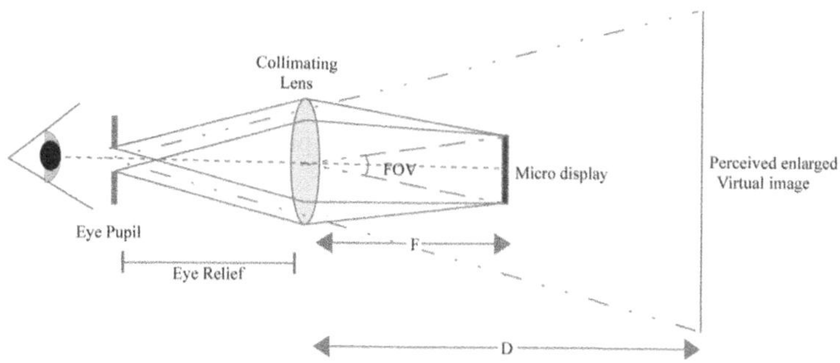

Figure 7.1. *Basic principle of near-to-eye optical system. This figure, and all other figures in this book, is available in full color at www.iste.co.uk/templier/OLED.zip*

The collimator lens is generally made of two or three lenses for low and medium FOV (<40 deg). The number of lenses is a result of optical design to maintain optical performances such as low distortion, low color artifact and acceptable modulation transfer function (MTF) to have sharp image and good pixel definition. For higher FOV, more lenses and more complex optical design is used with off-axis lenses and diffraction based aspheric surfaces. The design of the collimator system is mainly a function of the OLED size and requested FOV that determine the focal length and collimator footprint. The number of lenses and their nature (aspheric, conic, etc.) is a function of the OLED pixel size and exit pupil size. Generally speaking, injection molded plastic lenses are used for their rectangular lens format, low weight and a low production cost. However, plastic lens material type is limited to a few and this limits the choice of material to reduce optical aberrations. We also have to mention that some plastic materials, such as PCs, exhibits a large amount of birefringence, which adds more difficulty with a polarized near-to-eye see-through system. Finally, injection molded plastic lenses the advantage have a collimator lens system that can be preassembled together, which avoid of tolerance issues during near-to-eye system assembly. For low volume and professional use, it is preferable to use a glass lens to have higher image quality performances, but at higher cost. The transmission of the collimator lens system could be very high, as antireflection (AR) coating could be applied on each lens surface. A transmission ratio of >94% is possible even with plastic lens and standard AR coating for visible range.

7.2.2.1.1. Brightness considerations

We also have to mention that only a small solid angle emitted by an OLED pixel is useful for near-to-eye systems. The lightemitted by each pixel is Lambertian from an OLED display but only a small solid angle (generally <30 deg) reaches the exit pupil. Also, generally speaking, the light bundle from OLED microdisplay is not telecentric, which means that the chief ray angle from each pixel changes its direction over the OLED microdisplay surface. Since the OLED is emitting light and color in a Lambertian way, no difference is visible over the projected image. This is not the case for other kinds of microdisplay type (such as LCD or LCOS), where a specific backlight unit is needed. The design of this backlight unit has an important influence on color and brightness uniformity of the projected image.

The near-to-eye system as depicted in Figure 7.1 conserves the geometric extend (or commonly called system Etendue). The Etendue Law states that the product of pupil surface by the solid angle (or FOV) is constant [CAK 06]. The result of this fact is the conservation of brightness from microdisplay to the eye (almost, by taking into account the transmission Fresnel losses of the optical system). We will see that such a rule didn't apply for some see-through near-to-eye systems, where the exit pupil is formed by a specific exit pupil expander (EPE) that enlarges the user eye box but reduces the amount of brightness perceived by the eye compared to the OLED microdisplay.

7.2.2.1.2. OLED advantage for collimator design and image quality

Since OLED is an emitting display surface, there is no need to have a backlight unit as used for other microdisplay technologies (LCD, LCOS, etc.). Thus for

OLED-based systems, it is a great advantage to get rid of the backlight unit that increases the collimator footprint, and so to avoid heating, reducing the total power consumption of the system.

The reduction of collimator volume is also improved by the possibility to reduce the back focal length (BFL) for the collimating system (BFL is the distance from the microdisplay surface to the collimator first lens surface). This large distance is necessary when liquid crystal on silicon (LCOS) microdisplay type is used to fold the LED illumination light out of the backlight unit. For LCD, the backlight unit is behind the LCD panel, however, it still has a thickness of 3 to 5 mm.

There is also another advantage that is related to brightness uniformity of the projected image. Since OLED is a Lambertian surface and even if only a small amount of light out of the OLED reach the user pupil, the brightness and color uniformity of the projected image is always better, compared to LCD and LCOS microdisplays. We can also mention that for OLED, the contrast is higher and is independent of the viewing angle, which makes the contrast almost constant over the field of view and over the user viewing angle, which is not the case for LCD and LCOS.

7.2.2.1.3. Power consumption

The main advantage of OLED is its low-power consumption to generate the light directly, which means that it not need any external light source as it is the case for LCD, LCOS and DLP microdisplay technologies. For OLED, the light is generated in the form of juxtaposed RGB pixels. Native RGB color filter systems may have some advantage compared to LCOS sequential color for which a black and white display is used and sequentially illuminated by a sequence of red, green and blue from the backlight.

Table 7.1 shows some comparison of total power consumption of OLED microdisplay compared to color filter LCD (RGB dot pixels), color filter LCOS (RGB dot pixels), and sequential color LCOS (black and white pixels).

For sequential color LCOS, the frame frequency should be increased to beyond 360 Hz to avoid the flicker effect. A higher frame frequency is also recommended to reduce the color break-up phenomena (color breaks appears when white content moves quickly, or when the eye moves quickly). Color break-up also increases in peripheral vision, thus if the virtual image is not vision-centered the color break-up increases. To reduce the color break-up, frame frequency beyond 480 Hz, which increases the power consumption of the LCOS panel, is recommended. The color break-up phenomenon that is well-known for the DLP front projector is more critical for near-to-eye display, since the virtual image is always present in the user's eye. The higher the field of view and the display brightness, the higher the color break-up phenomenon. Figure 7.2 shows an illustration of the color break-up phenomena, as it is seen by the user.

Microdisplay technology	OLED	LCD (color filter)	LCOS (color filter)	LCOS (Color sequential)
Panel power consumption [1] (mw)	50	50	40	150
Controller power consumption (mw)	100	150	150	150
Backlight power consumption[2](mw)	0	150	200	250
Total power consumption (mw)	150	350	390	550

Table 7.1. *Average power consumption comparison: OLED versus LCD, color filter LCOS and sequential color LCOS*

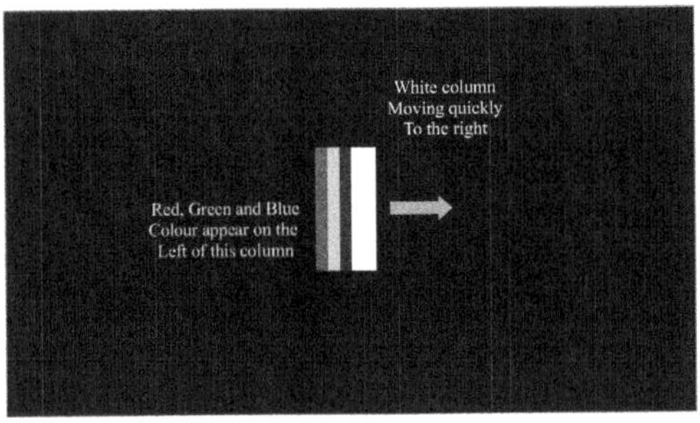

Figure 7.2. *Illustration of color break-up resulting from sequential color functionality for near-to-eye*

7.2.2.1.4. Brightness efficiency

If OLED microdisplays show a lot of advantages power consumption, reduction of system footprint and better image quality, the current OLED displays in terms of do not have the high brightness required for near-to-eye systems and specifically for see-through devices, where the brightness of the projected image has to compete with the outdoor environment. Brightness efficiency defined as the ratio between the user perceived brightness over the native display. Brightness depends on the nature of the optical system (polarized or not) and on the type of the see-through used.

1 Calculated for <1000 Cd/m² brightness at microdisplay level.
2 For WVGA/VGA panel.

Among the see-through systems, some are with or without exit pupil expander along with light guide based technologies that will be described in the following paragraphs. Exit pupil expander brightness efficiency is in the order of 10 to 25%.

Outdoor scene average brightness is about 6000 Cd/m^2 for a sunny day. It is necessary then to have an image brightness of 3000 Cd/m^2 in order for the user to still see a decent contrast for the projected information into the eye. It is possible to increase the perceived brightness versus outdoor brightness by adding a sunglasses to the near-to-eye device. Current sunglasses photopic transmission ratio goes from 3 to 8% (grade 4), 8 to 18% (grade 3), 18 to 43% (grade 2), 43 to 80% (grade1) to >80% (grade 0), which allow an image brightness of 500 Cd/m^2 to still be acceptable. However, adding low-transmission sunglasses is not an ergonomic solution, since for better communication, it is important to see the user's eye and not to hide them from an audience.

Another way to avoid dark sunglasses is the use of a photochromic lens in front of the near-to-eye display. Photochromic lenses change their transmission when exposed to UV sun radiation to become darker, while fully transparent in indoor environments. Other than the slow response time of these lenses, their costs are also high and limit their use on a large scale.

One way to adapt image brightness to the outdoor scene is to implement a light sensor inside the near-to-eye system so that image brightness could be adjusted automatically.

It is then mandatory to improve both the see-through brightness efficiency and also the native brightness of OLED microdisplays. The first point will be addressed in the see-through display technologies section ahead. The improvement of OLED native brightness is currently the subject of heavy R&D effort from main OLED microdisplay suppliers. Chapter 8 of this book addresses the current development of high brightness OLEDs.

Microdisplay-based collimator systems, as described in Figure 7.1, cover the majority of optical systems used in the HMD industry. However, after the light is collimated, the way to carry the collimated image in front of the eye uses different optical elements that yield a variety of HMD systems; the obstructed (or non-see-through) systems, and the true see-through (or transparent) systems.

7.2.2.2. *Obstructive non-see-through optical systems*

This kind of near-to-eye system displays a virtual image into the user's eye, by fully obstructing the outdoor environment of the user. Several optical systems exist. The simplest is to have the collimator directly in front of the eye, others uses a 45° mirror in order to position the collimator system on the side or on the top of the user's eye. A good description of existing systems is summarized in the reference [CAK 06]. Several non-see-through consumer products have been in existence for many years, but did not get their place in the consumer and professional market. The reason relates on the inappropriate blocking of vision for the user for both entertainment and professional applications. For the consumer, blocking the user vision implies that the user could not be in a mobile world, then the user is limited to be at home, where he already has a full HD large screen to see videos, personal computer (CP) content and to enjoy internet content. For professional use, where awareness of user environment is a key factor, the obstruction of user eyes is a show stopper.

Several existing non-see-through products are of binocular type. They are used to see video content generally provided by wired connection to a local multimedia player or to a smartphone. These products are able to display true 3D (stereoscopic) video, since right and left video channel are separated physically. Thus, these products are adapted to virtual reality applications, like 3D games, flight simulator and 3D applications for product design and training process. Currently, two non-see-through binocular products use OLED technology (Sony and Carl Zeiss). The current old brightness (~300 cd/m^2) is enough to display 3D video with obstructed vision. The current products have very good image quality and especially a very good contrast (Figure 7.3).

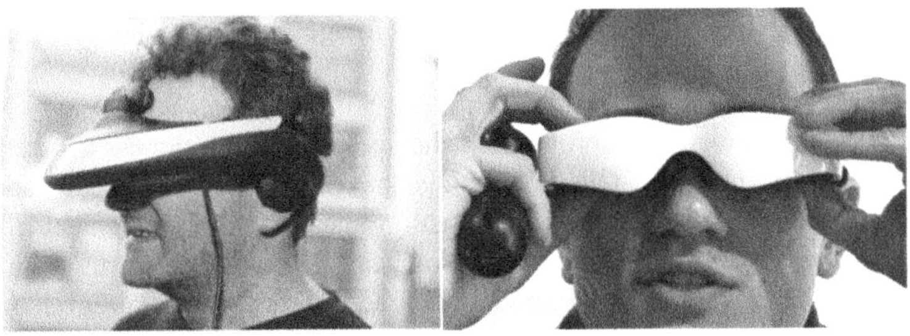

a) b)

Figure 7.3. *Non-see-through OLED-based HMD: a) by Sony and b) Carl Zeiss*

Some binocular products adapted to gamers recently received a lot of interest with a large virtual image (FOV~110 deg) to create a fully immersive 3D game environment (Figure 7.4). The idea here is to use an available direct view display with 1280 × 480 pixels and a pair of optics to enlarge half of this image by 640 × 480 pixels for each eye.

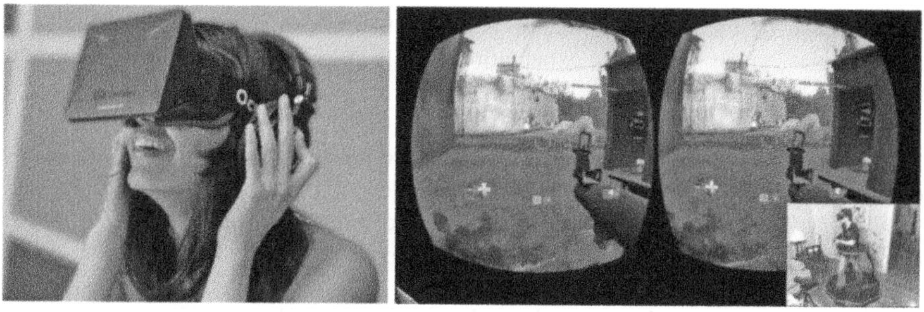

a) b)

Figure 7.4. *a) Non-see-through glasses with 110 deg FOV for game application, and b) a view on the two images generated by one direct view display with one pair of optics and an image processing algorithm*

Recently, several obstructing monocular products have emerged that address the consumer market (mainly sport) or for the professional market (logistic, warehouse, etc.). The display (small, FOV~16 deg) is generally located at the vision periphery (Figure 7.5). The user must move his eye to focus on the display and then move it again to see the environment. This is not a very comfortable situation for the user, moreover, eyestrain occurs since the right and the left eye do not have the same content and same average brightness. In fact, display brightness for non-see-through systems is very low (<400 cd/m^2), since display brightness does not compete with outdoor scene brightness. We have to notice that non-see-through display could not be used for augmented reality application, since the display image is isolated from the outdoor scene. We can consider non-see-through display for near-to-eye products as a display to look at from time to time obtain the information we may need, as we do today with any other display systems (smartphone, tablet, PC, etc.).

Figure 7.5. *Non-see-through monocular wireless HMD products. The LCD-based display unit is placed out of the center vision to be used in glance mode*

We have to mention also that it is easier to build a non-see-through OLED microdisplay, since brightness expectation is still low and compatible with current brightness levels.

As a conclusion about obstructive, non-see-through product, two categories emerge: binoculars with large FOV for immersive virtual reality gaming and simulation. The other category is for mobile and industrial use. We have to notice that current products are still far from customer expectation.

The trend now is to have see-through or transparent display, since the near-to-eye system is used in mobility, where information and video content should be available to the user at any time, anywhere and should be superimposed on the user outdoor scene.

7.2.2.3. *See-through optical systems*

See-through near-to-eye display had been the focus of military and more specifically avionics applications for more than one decade. The target of several developments was to display critical information into the eye of the pilot, always in situation awareness during their missions. Two kinds of near-to-eye systems emerged; the first uses the semi-reflective curved pilot visor as a part of the optical system, the second uses a standalone display system that is fixed to the pilot helmet. Recently, in the consumer domain, several start-ups at one major internet company announced the development of see-through monocular products, known as "smart glasses" or "AR glasses".

In the following sections, we will describe the basic concepts of optical see-through technologies and detail how OLED microdisplay technology could fit within their future development.

7.2.2.3.1. Overview of existing technologies

Various techniques have existed for some time for see-through video wearable displays. Most of these techniques can be summarized into two main families: "curved-mirror" based and "waveguide" based. The curved-mirror based techniques use semi-reflective curved mirrors placed in front of the eye with an off-axis optical projection system.

7.2.2.3.2. Curved-mirror based see-through near-to-eye system

Figure 7.6 shows the basic concept for curved mirror near-to-eye optical system, defined as pupil forming system. The system depicted in Figure 7.6 is the basis of the Olympus patent [HOS 96]. The light out of the microdisplay is collimated and an intermediate pupil is formed within the optical system. The curved mirror conjugates the external pupil from the collimated optics into the user's eye pupil. The curved mirror placed in front of the user's eye has a semi-reflective coating to reflect part of the light from the collimator and to allow the outdoor scene to be transmitted to the eye.

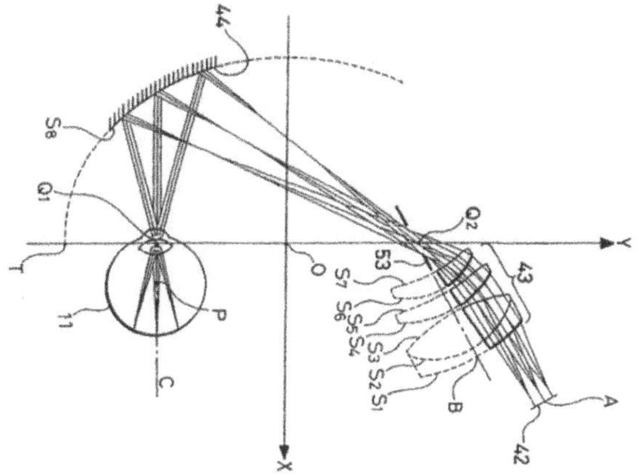

Figure 7.6. *Basic principle of the curved-mirror near-to-eye optical system [HOS 96]. The microdisplay image is enlarged by an optical collimator that forms an intermediate pupil. The curved mirror conjugates this pupil into the user's pupil*

The curved mirror technique suffers from a high amount of distortion, which needs to be corrected optically or electronically adding cost and reducing image resolution. Some attempts use a free form curved mirror, in order to reduce the amount of distortion, about 30% for the original system. Moreover, certain implementations have a small "eye motion box", which is the equivalent of looking through a keyhole to see the image. This is uncomfortable for the user and requires mechanical adjustment, further adding to cost. The major issue comes from the form factor, which is not appealing and handicaps end user adoption (see Figure 7.7).

Figure 7.7 shows two implementations of this technique for a near-to-eye systems in the form of glasses; the other technique is using the helmet visor itself, acting as a semi-reflecting mirror mounted in a pilot helmet. We have to mention that semi-reflecting coating uses a notch filter (one transmission narrow band wavelength) for monochrome display, or a metallic semi-reflector coating to reflect white light.

Recently, a holographic curved reflector was used. These reflectors are generally used for monochrome displays. The curved surface is a holographic optical element (HOE) that plays the role of curved mirror for the recorder wavelength. The problem with HOE components is their low efficiency to reflect larger wavelength bandwidth, since the HOE does not use Bragg conditions, as the incident light is in the air. We have also to mention another limitation related to the difficulty to deliver large eye boxes without color and geometrical aberration. This is why HOE reflectors are used for low FOV systems (<15 deg), generally with a monochrome display.

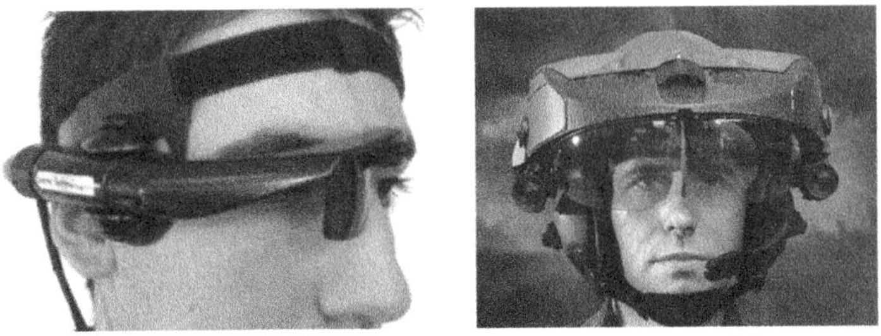

a) b)

Figure 7.7. *Curved mirror based see-through HMD. Examples of a) standalone system in the form of glasses, and b) embedded system on helmet*

7.2.2.3.3. Light-guide based see-through near-to-eye systems

The second family of see-through display is the so called "light-guide" or "waveguide" based techniques. These present the most promising technologies for wearable displays since they reduce the cumbersome display optics and electronics in front of the user's face and in the user's line of sight. Using a waveguide, the physical display and electronics can be moved to the side (near the user's temples) and a fully unobstructed view of the world can be achieved, therefore opening up the possibilities to true augmented reality experiences. Various waveguide techniques have existed for some time for see-through wearable displays. These techniques include diffraction optics, holographic optics, polarized optics and reflective optics.

7.2.2.3.3.1. Diffractive waveguide

The diffractive techniques use deep slanted diffraction gratings (i.e. Nokia technique now licensed to Vuzix). This technique uses slanted gratings to in-couple collimated light entering the waveguide at a particular angle, then, the light travels through the waveguide using the principle of total internal reflection or "TIR", and finally, the light is extracted to the eye with another set of slanted gratings (Figure 7.8). One of the first publications describing this technique is from Ann Arbor University [UPA 85]. The schematic for this technique can be seen in Figure 7.8, as published by Nokia [LEV 07].

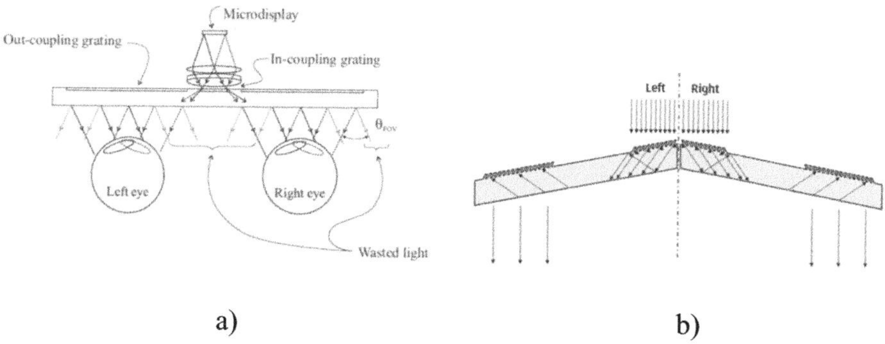

Figure 7.8. *Schematic of slanted grating technique claimed by Nokia. a) The binocular system uses a single LCD microdisplay, b) the detail shape of slanted deep grating*

The basic principle as stated in the Ann Arbor patent publication, dated in July 1985 [UPA 85] stated that in-coupling grating and out-coupling grating should be symmetric to minimize the optical aberrations. The basic drawing from the original patent is depicted in Figure 7.9.

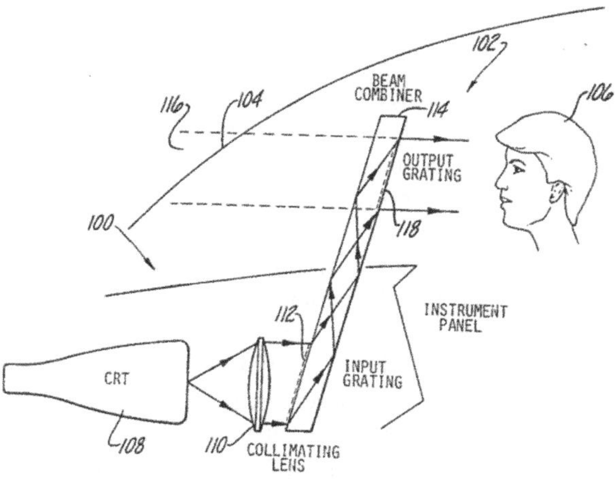

Figure 7.9. *Basic principle of grating based see-through display as patent by Ann Arbor University*

The slanted feature added by Nokia is to increase the diffraction efficiency of the grating for the visible range. The current structure is nearly 150 nm with a grating period of the same amount.

Figure 7.10 shows two near-to-eye systems using the diffraction grating technique. Nokia prototype never reach the market, however, Vuzix who get a license from Nokia announced a monocular version for professional use.

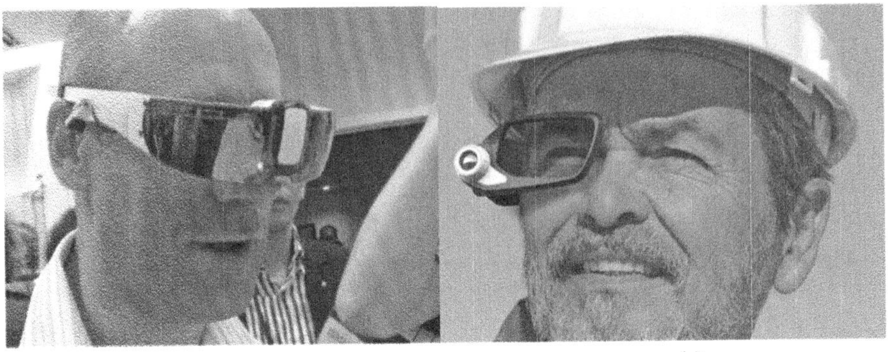

a) b)

Figure 7.10. *See-through diffraction grating based HMD prototype by a) Noika and b) Vuzix M2000AR eyewear product*

The diffraction grating technique presents some key challenges. The first is producing the deep and slanted nanometric grating structures at low cost. The technique for producing these deep slanted structures is not something that is commonplace today in traditional optical component manufacturing. Therefore, the technique will remain costly for some time. The second issue with this technique is that it produces color non-uniformity in the image. Since light is in-coupled and out-coupled at a certain angle when it hits the diffraction structure, it creates a "rainbow effect" due to the variation of spectral reflectivity versus the incident angle within the image. This means that the various reflected wavelengths do not have the same amplitude when they encounter the diffraction pattern at an angle. The diffractive technique therefore works best with monochrome-based systems but this is a big limitation for the consumer space where full color is extremely important. The third aspect is that this technology is intrinsically limited in the field of view (FOV). Large FOV displays (large virtual screens) are not possible using this technique again due to the variation of spectral reflectivity vs. angle issue. The higher the incidence angle, the higher the color non-uniformity. Higher angles are needed for higher FOVs and if the FOV is increased beyond 20°, the color non-uniformity becomes very noticeable as the human eye is extremely sensitive the color non-uniformity variations.

We have to mention finally that this technique offers a good see-through capability and large eye box, along with a thin-light guide of <2 mm. Still the improvement of color issue and limited FOV seems to be a big physical limitation for the technique, especially in the consumer space.

When used with an OLED microdisplay, the diffraction grating technique reduces the total brightness efficiency, since light should be polarized to be diffracted efficiently with the grating, which leads to an additional loss of more than 50% of OLED native brightness.

7.2.2.3.3.2. Holographic waveguide

The holographic technique is quite close to the diffraction grating technique described above with the exception that a holographic element is used to diffract the light [MUK 08]. Holograms work by reflecting certain wavelengths of light. In this way, the incident light is reflected at a certain angle with regard to the hologram. Holograms are intrinsically limited when used in a waveguide due to the fact that the reflected light loses intensity with angular variation. Only limited angles are possible in order not to lose too much light and to keep good image uniformity. Therefore, this technique is intrinsically limited in FOV. This technique is also plagued by color issues known as the "rainbow effect". Holographic elements reflect only one wavelength of light so for full color, three holograms are necessary; one that reflects red, green, and blue, respectively. This not only adds cost, but since the three holograms need to be "sandwiched" together, each wavelength of the light is slightly diffracted by the other color hologram adding color "cross-talk" in the image. Therefore, the eye sees some color non-uniformity or color bleeding when viewing the virtual image. Some of this color non-uniformity can be corrected electronically but there are limits to this, as the human eye is extremely sensitive to this phenomenon. The holographic technique is used in the Sony and Konica Minolta as shown in Figure 7.11.

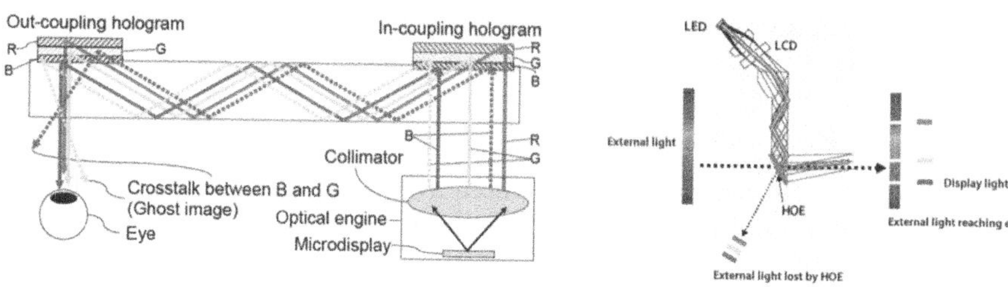

a) b)

Figure 7.11. *Holographic see-through HMD from a a) Sony, b) Konica-Minolta concept*

Another method is using only one hologram with monochrome light. This method is used by Sony for a pair of glasses to display text subtitle for cinema. The other application is used by BAE systems in the military for avionics applications.

Figure 7.12 shows photography of Sony prototype and Q-sight product from BAE system. The latter is a monochrome monocular type, dedicated to helicopter equipment integrated into the pilot helmet.

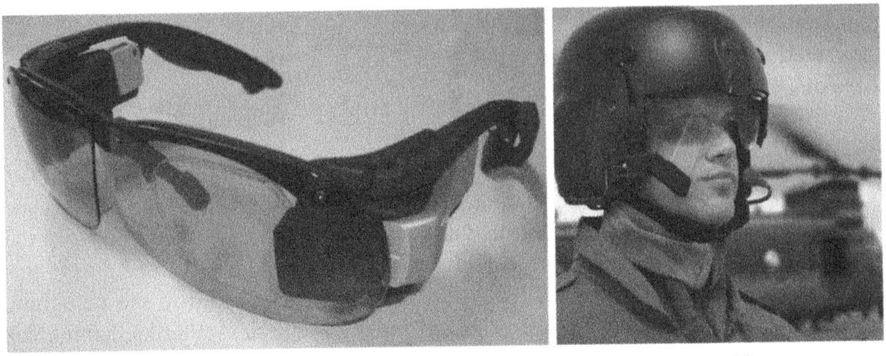

a) b)

Figure 7.12. *a) See-through HOE based HMD prototype by Sony and b) Q-sight military product by BAE systems*

The fact that optical collimator is located far away from the extraction area (in front of the user eye) and the fact that light guide is so thin (<2 mm), a large number of reflections occur. For each reflection, only a very small part of the image is extracted. This leads to a very poor efficiency at the cost of large eye box. Nevertheless, we have to mention that the holographic technique offers the best see-through transmission ratio (~85%).

The diffraction efficiency of the hologram is maximized for one polarization. As a result the total brightness efficiency with an OLED microdisplay is reduced by a factor larger than 50%, since OLEDs emit a non-polarized light.

7.2.2.3.3.3. Polarized waveguide

The polarized waveguide technique is used by Lumus [PCT 06]. This technique uses multilayer coatings and embedded polarized reflectors in order to extract the light toward the pupil. The light out of a collimator is in-coupled into a glass-based light guide (of approximately 2 mm thick) by an embedded reflective polarized coating. The light then hits an array of reflectors embedded into the light guide substrate. The light guide is made by an assembly of several parts that embed three or six reflectors. The reflectivity of each reflector is calculated to have 1/n (n is the number of reflectors). But since the 1/n rule could not allow the image to be uniform, each reflector has a reflectivity that increases from the first reflector to the last reflector. However, since the polarized coating and its angular sensitivity is difficult to obtain, an average reflectivity from 20 to 25% is claimed from the first reflector to the last one for three reflectors or ~10% for six reflectors. The multilayer polarized coating should also have an angular sensitivity to transmit the image angle at a high-incidence angle and to reflect the image angle at a low-incidence angle. This technology does not suffer from the small FOV issues and the eye motion box can be quite large. However, it suffers from several major drawbacks that do not allow a viable solution for a low cost, consumer-based solution. The polarized coatings are multilayer coatings of 25–30 layers each and must be deposited on glass, as plastic is not compatible with this high-temperature process. Several of these reflectors need to be glued together precisely with extremely tight tolerances on parallelism, cut at an angle (again, with a very high level of parallelism), and polished in order to make the waveguide. Each reflector needs to have a different amount of coatings ranging from 25 to 30 layers in order for the virtual image to be uniform. This process is not geared toward high-volume and low-cost manufacturing. Furthermore, glass is fragile and the perception of a thin piece of glass in front of the eye will be detrimental to large scale user adoption. Additional drawbacks come from the fact that the system and reflectors are polarized, making the system inherently inefficient and mainly when using an OLED display, because nearly 60% of the light is lost when it is reflected. Moreover, the "rainbow effect" of color non-uniformity exists due to the polarization states. Figure 7.13 illustrates this technique and the Lumus prototype.

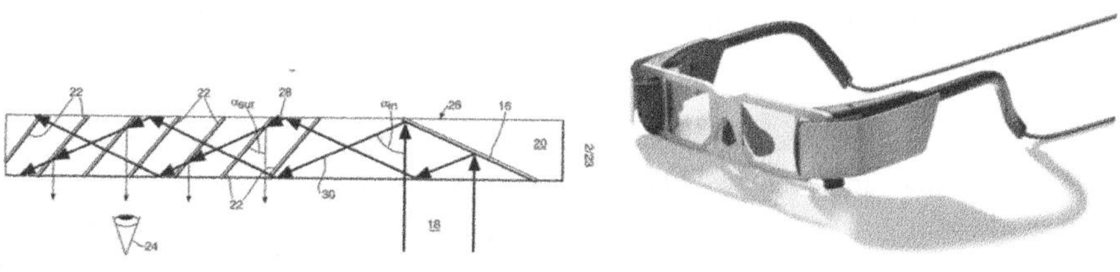

a) b)

Figure 7.13. *a) The schematic from the Lumus patent and b) the Lumus prototype*

The use of a glass-based light guide in front of the eye raises a legitimate question about the use of such glass technology in consumer spaces. In fact, the American FDA impact resistance lens test [x] (FDA-2007-D-0367) states that the impact test of 5/8 inch steel ball weighting approximately 0,56 oz and dropped from a height of 50 in from the horizontal upper lens surface should not break the lens (here the light guide). This rule is to be applied of course to any light guide based technology that intends to address the consumer market and mainly in the USA.

7.2.2.3.3.4. Reflective waveguide

The reflective technologies have the advantage of using straight reflective optical components (no grating or HOE components or multilayer coatings). They do not suffer from the color non-uniformity issues since they use "normal" semi-reflective mirrors, therefore reflecting white light without any degradation. The possibility to use a molded plastic substrate for the light guide is also a key advantage of this technique. This type of system is used by Epson in their Moverio product, the Google Glass prototypes, and by Optinvent in the Clear-Vu technology. The approach taken by both Epson and Google uses a single reflector embedded into the light guide (although Google implementation does not use TIR). Other than the lack of innovation, and therefore a low entry barrier, the problem with this approach is that the size of the reflector is directly proportional to the FOV (field of view) and eye motion box dimension; therefore the light guide becomes quite thick. In both the Google and Epson cases, the light guide thickness is around 1cm as seen in the figures below.

7.2.2.3.3.5. Single-reflector waveguide method

Two wedge-shaped pieces are glued together precisely (one is coated with a semi-reflective coating and the other is bare) in order to produce the waveguide. These wedge shaped structures need to be molded with high precision molding techniques. This is the main challenge of this technique. As with the other waveguide technologies, an optical collimator magnifies the image generated by a microdisplay and injects it into the light guide.

Currently two products are using this method. Both have a limited FOV and thick waveguide. In fact, the FOV is directly proportional to the thickness of the light guide.

The first product is from Epson. The collimated light is in-coupled into a thick waveguide of 1 cm thick. Through the TIR principle (total internal reflection), the light travels through the light guide and is extracted using one semi-reflective mirror. Epson is using transmission RGB dot panel inside the Moverio product. The product is using binolcular systems that could allow see-through 3D video. However, the current product is quite heavy (~210 g), which limit is its use in the consumer space. Figure 7.14 shows the principle of this technique and the Moverio product.

The other single-reflective mirror is used by Google for their Glass explorer version product. The light out of the sequential color Lcos microdisplay is not collimated prior to being in-coupled to the light guide, which has a big advantage of saving space (Figure 7.15). The light crosses through a 45 degrees semi-reflective mirror and hits a reflective aspherical mirror that collimates and magnifies the microdisplay image. The collimated light hits the 45 degrees semi-reflective mirror and back is extracted toward the user's eye. The light is not transmitted by total internal reflection, but crosses through the thick plastic light guide (of ~1 cm). As light is transmitted and reflected by a metallic semi-reflective coating, the total efficiency is low.

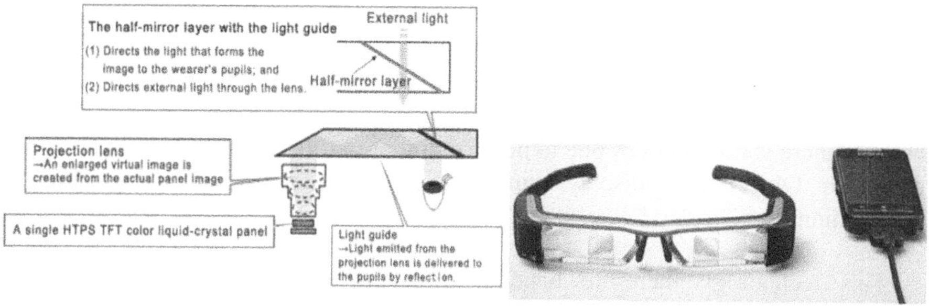

Figure 7.14. *Basic principle of Epson Moviero product seen on the right*

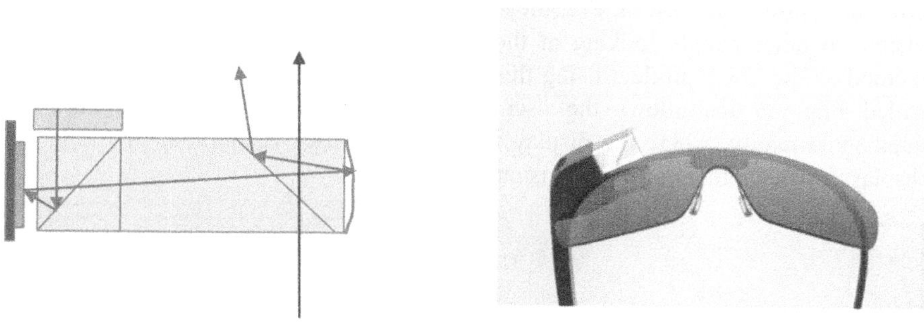

Figure 7.15. *Basic principle of Google Glass product seen on the right*

7.2.2.3.3.6. Surface array reflector waveguide method

This method used by Optinvent is fundamentally different than the above. It uses a surface structure made up of several reflecting mirrors, which make it possible to have a thinner light guide while maintaining a large eye motion box as well as a large FOV. This surface structure allows Optinvent to mold a monolithic light guide (out of one piece of plastic), which is coated with a semi-reflective coating [SAR 13]. A "cover plate" is glued to this piece of plastic in order to protect the structure and to assure the optical see-through function. This cover plate component does not need to be precise since it is not used to generate the virtual image. It only assures the see-through function by compensating the prismatic effect when the pupil looks through the structure of the light guide.

The mirrored structure using traditional coatings found throughout the optics industry. This will allow the components to be made using traditional coating techniques, therefore reducing cost. Consequently, any type of microdisplay can be used in these systems since there is no polarization required (LCD, LCOS, OLED). This reflective system also tends to be more efficient in power consumption, because there is no light loss due to polarization or grating/holographic effects when using an OLED microdisplay, since reflective mirror array is agnostic to polarization.

This technology therefore benefits from all the advantages of reflective waveguide techniques (no color issues, molded plastic substrate, traditional coatings, better efficiency, large eye box and FOV). Moreover, it has the additional benefits of a thinner waveguide made out of one monolithic piece of plastic, therefore improving the form factor and further reducing cost. The main challenge of this technology is to mold the light guide and its surface structure precisely and to design a system that finds the right compromise for performance and cost. Since mirror array are close to the user's eye, they are not seen by the user, however, they are visible to other people looking at the user. Figure 7.16 details the mirror array method as the ORA product using this method. This product shows a new feature called Flip-Vu that allows the user to use the display in two positions: the augmented reality, where the display is centered to user vision and, by tilting the display down, another dashboard vision is possible.

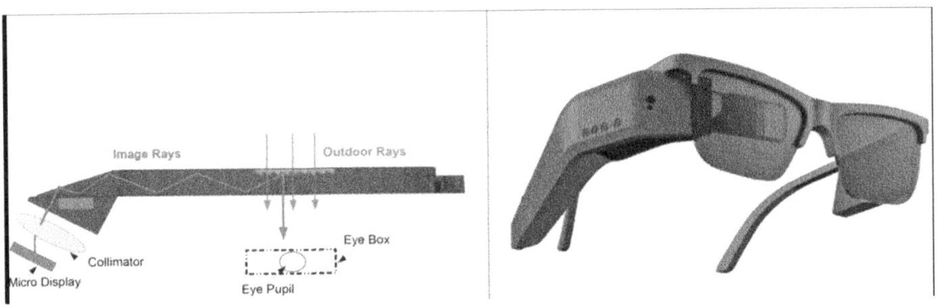

a) b)

Figure 7.16. *a) Basic principle of Optinvent mirror array method, and b) the ORA product*

7.2.2.4. *Conclusions about near-to-eye optical system*

Table 7.2 summaries the different solutions for near-to-eye systems.

	Light guide Based See-through Near to Eye technologies				
	Diffraction technique		Non Diffraction Technique		
Principle	*Holographic Optical Element*	*Deep Diffraction Grating*	*Embedded Polarizing reflector*	*Array of small mirror*	One semi-reflective mirorr
Schematic	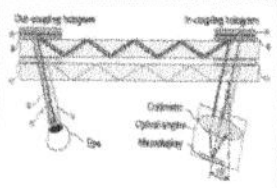		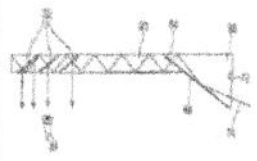	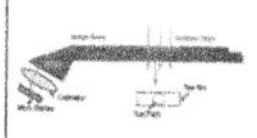	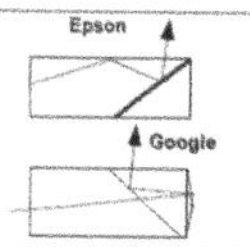
FOV (Deg)	<20	<20	<32	24	14 deg (Google), 23deg (Epson)
Production	Non Standard Exotic component	Non Standard Exotic component, IC wafer based process	Impossible to produce in large scale	Standard existing production	Standard existing production
Cost	High	High	Very High	Low	Low
Colour issue	Yes (rainbow effect and white color non uniformity)	Yes (rainbow effect and white color non uniformity)	Yes (residual PBS color)	No	No
See-through Ratio	85%	~70%	75%	50%	~40%
Light guide weight	Two glass guides (2x1.4mm) Sony	Glass (2mm),Nokia (licence to Vuzix)	Glass (2mm), Lumus	Plastic (4mm), Optinvent	Thick plastic (10mm), Google/Epson

Table 7.2. *Summary of different see-through techniques*

Among the products and method discussed for obstructed vision near-to-eye systems, we confirm that for augmented reality and informative applications, current products are transition products and the future is for transparent display that are more adequate for mobile use. However, for 3D gaming or virtual reality applications, obstructed near-to-eye systems should continue to offer large FOV and higher resolution with a better form factor for the glasses.

Of the various methods discussed for transparent (see-through) near-to-eye systems, waveguide technologies with reflective type seems to be the most promising for large scale consumer deployment. The main advantages are the lower cost, plastic substrate and lack of color issues. The OLED panel is then more adequate to be used by reflective methods, since light out of OLEDs is not polarized, thus more than 60% of light is saved.

7.2.3. *Evolution and challenges for near-to-eye wearable display systems*

The main challenges for near-to-eye display can be separated in those related to the display and the others related to the system in general.

For the system in general, the trends are as follows:

– Reduce system volume:

- with the display: being self-emitting display, the OLED helps;

- with the battery: together with higher specific capacity, they will have to be "any shape" (including curved) to be easily integrated in any new compact system;

- with the electronics: a size reduction of the growing number of needed ICs: controller, memory, position sensor, camera, GPS/WiFi/BT, and associated boards.

– Reduce power consumption:

- of all the above-mentioned devices;

- most of IC manufacturers (Intel, Qualcomm, ARM, etc.) work hard toward this objective, to find new dedicated IC technologies for wearable devices.

Concerning the display, the challenges are, in the order of priority:

– higher brightness;

– maintain lower power consumption;

– extend color fidelity and lifetime for both consumer and professional applications.

High brightness is requested, since in see-through near-to-eye systems, perceived brightness by the user is about 20% of the native OLED brightness. Then a minimum 5000 Cd/m^2 is required at the microdisplay level, where the current brightness level is 10 times lower.

Color fidelity is required for immersive application, where video quality is a key factor. Lifetime is an important factor to address for OLED panels, since it is quite a recent technology.

Lifetime of more than 20,000 hours is required for consumer products and even more for some professional applications.

Several technical solutions should be able to address these issues, as it will be described in Chapter 8.

7.3. Electronic viewfinder embedded into a camera/camcorder

For this other important application of OLED microdisplays, the following aspects will be addressed:

– general requirements for viewfinder applications;

– optical system;

– evolution of viewfinders.

7.3.1. *Introduction and general requirements*

Over the past decades, the optical viewfinder dominated the camera market industry. The optical viewfinder consists of lenses and a pentaprism (Figure 7.17 [TAK 14]). However, within the last decade, the upcoming mini display technologies started to enable the first direct view displays onto camera bodies. Due to reflections and brightness limitations at ambient light conditions these displays where often not useful as the user needs to switch back to the optical viewfinder.

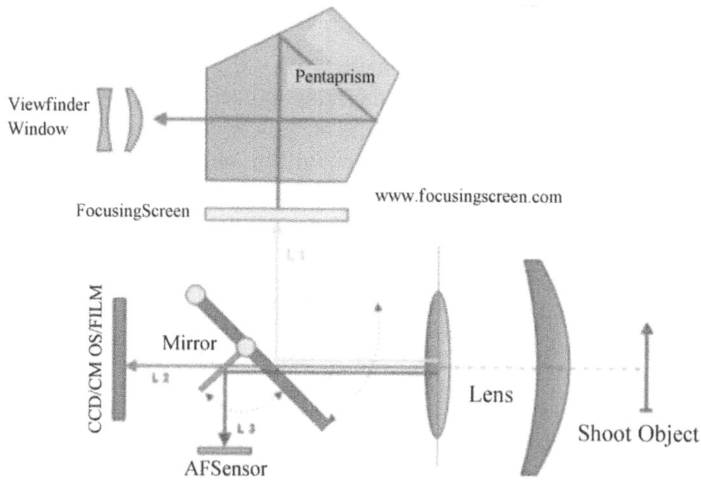

Figure 7.17. *Optical viewfinder*

Nowadays, the optical camera viewfinder changes to the electronic viewfinder (EVF) in consumer and professional cameras. The job of this EVF is to display the ambient, which is captured by the image sensor into the user's eye via a microdisplay. This enables the complete digitalization of the capturing-imaging chain in cameras. To reach acceptance the viewing action of the EVF has to be similar to the optical viewfinder.

The following are advantages of electronic viewfinders over optical viewfinders:

– increased available space and decrease weight of cameras due to absence of optical elements;

– shows immediately effects by changing exposure, focus and f number;

– provides a more accurate depiction of the contrast, white balance and color saturation that will appear in the final image.

– no parallax driven by using the same chain as the capturing unit;

– additional information can be overlaid like a tilt sensor;

– 100% coverage of the final image.

Like all technologies, electronic viewfinders also have their drawbacks:

– power consumption and therefore operating time gets smaller;

– fast movement can lead to latency and therefore to artifacts.

Nowadays, high-end EVF are dominated by OLED displays, which are caused by the very good contrast ratio and their low-power consumption compared to LCOS and LCD technologies. To overcome the power consumption of an additional display, most camera manufacturers have built-in sensors. With this setup, the camera is able to detect the viewing mode and then switch to direct view mode or to EVF. Table 7.3 shows some available EVF.

	Fujifilm X-T1	Panasonic Lumix DMC-GH3	Sony Alpha 77	Sony Alpha 57	Mountable Olympus VF-5
Display type	OLED	OLED	OLED	LCD	OLED
Aspect ratio	4:3	4:3	4:3	4:3	4:3
Resolution of display [x:y]	XGA	XGA	XGA	SVGA	XGA
Maginification	0,77	1,34	1,09	1,04	1,48

Table 7.3. *EVF overview*

7.3.2. *Optics*

In general, an EVF consists of common eyepiece designs with a short focal length. Instead of an intermediate image, the display is placed, to magnify and project into the human eye. Eyepieces for EVF consist of two–five lenses to correct optical aberrations and enable a bigger FOV. Figure 7.18 shows the geometry of a magnifier [TAK 14].

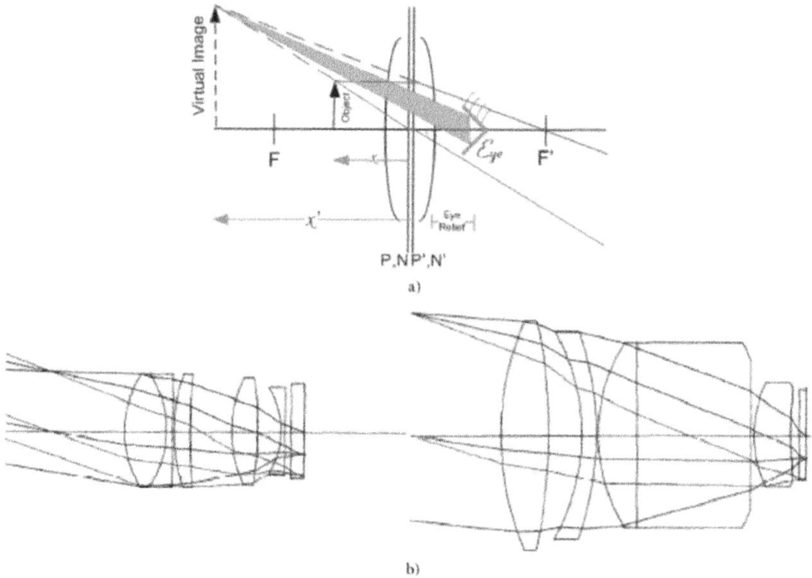

Figure 7.18. *Geometry of magnifiers*

Figures 7.19, Figure 7.20 and Figure 7.21 give an overview of well-known eyepieces. These theoretical eyepieces are then adapted to the specific requirements of the microdisplay.

Orthoscopic eyepiece

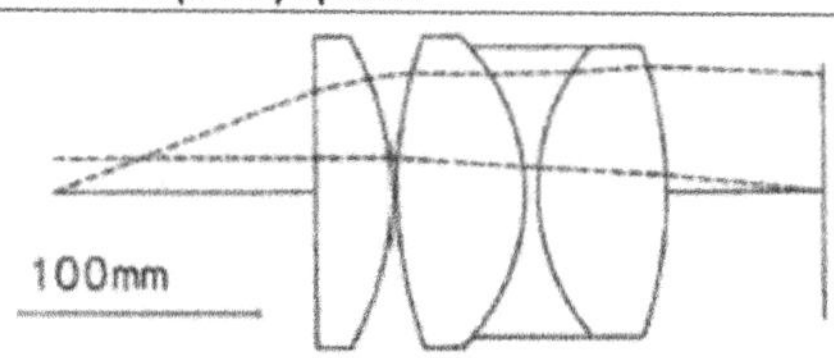

- plano-convex element
- 20 to 25 degree HFOV
- Noted for long eye relief
- Known for good distortion correction
- Field coverage is then limited by HO astigmatism

Application: low end binoculars

Figure 7.19. *Orthoscopic eyepiece [CAK 06]*

Plossl

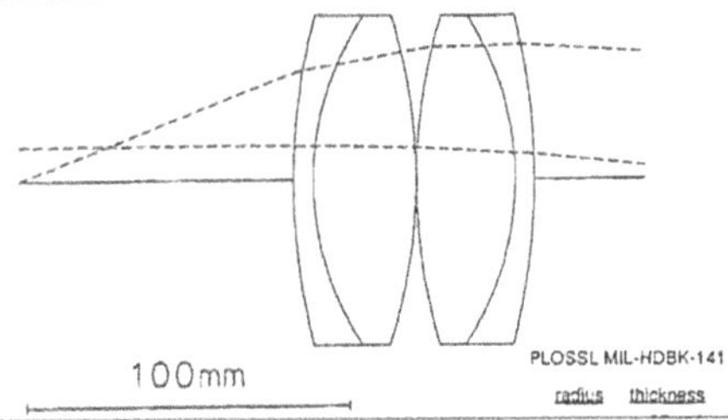

- Long eye relief
- Two doublets
- Better overall performance than orthoscopic with the exception of distortion
- Versatile, general purpose, medium field (25 degree) eyepiece

Application: Military

Figure 7.20. *Plossl eyepieces*

Kellner eyepiece

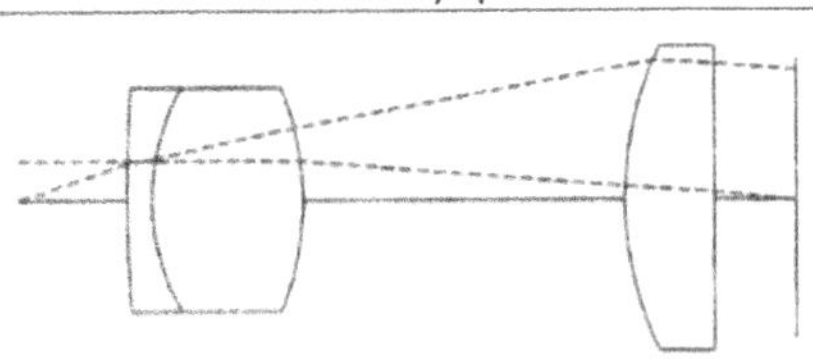

- Improves the color correction, in particular lateral color
- !arger field of view
- The index difference in the doublet is important for astigmatism control

Application: astronomical observing

Figure 7.21. *Kellner eyepiece*

The following paragraph lists all the relevant parameters with corresponding values.

The field of view is one of the main parameters for an EFV optic design concept. The angle of view can be measured horizontally, vertically, or diagonally. Common available systems start from 20 up to 40 degrees in diagonal with an aspect ratio of 4:3. This allows EVF to show 100% of the viewing/capturing field. In general, an angular resolution of 1 arc minutes/pixel allows resolution, which imitates the resolution of a human eye. The magnification of such designs variegates from 0.7 to 1.5. An additional large eyebox (> 6 mm) is required to enable the pupil movement without large vignetting. An eye relief of >15 mm, which is the distance of the last lens surface to the pupil entrance, is required for a comfort viewing zone. Due to the fact that the image is projected by a plane surface, the field curvature has to be corrected. This enables the user to view a sharp image all over the field. Even when the image is sharp, the image distortion can be easily detected at the edges of the scene, which also has to be taken in account.

Other than the optical performance, human perception also has to be considered. Therefore, the display characteristics play an important role. The minimum requirement for resolution of the display starts at SVGA (800 × 600 pixel) with a color depth of 8 bit per channel. The color space, which is often not mentioned, is also important for the color perception and should be NTSC. This is a value for the deviation from the primary color and the white point.

The display of an EVF provides enough luminance to view the scene in bright outdoor conditions. In fact, this is an eyepiece-based system, so brightness is not an issue at all. However, common microdisplays provide from 100 to 500 cd/m^2, which fulfill these requirements. Another critical parameter is the variations of the homogeneity over the image field. The frame rate and the frequency drive the latency of the display. During fast movement, these artifacts can be recognized. To overcome this issue, frame rates of 60 Hz per channel are recommended.

Nowadays, all EVFs offer a diopter adjustment. This is done by turning a wheel next to the EVF. Turning the wheel leads to changing the distance between display and lenses by a mechanical stage. With this, prescription glass users are able to view without correction glasses.

7.3.3. *Evolution of view finders*

The increase of cameras equipped with EVF shows a trend in this direction. Actual trends for microdisplay like increased brightness and higher resolution will also affect the EVF in the future. Because of this, HD EVF will reach the market soon.

For 3D video cameras, more review and more autostereocopic displays are of interest. The need is to directly review 3D content during capturing. This could be easily solved by a binocular approach. Hereby two separate channels (left eye, right eye) have to be shown to the users' eyepiece. With this, real 3D videos can be directly viewed during capturing.

7.4. Other display systems with OLED microdisplays

As explained in the introduction, brightness of OLED microdisplays is limited to a few 100 Cd/cm^2 (color); therefore such displays cannot address applications such as head-up display, and normally even less projection systems. However, as we will see first, some attempts to develop picoprojectors using OLED microdisplays have been made.

However, the core domain of OLED microdisplays is clearly near-to-eye applications, and their main objective will be to extend from immersive (HMD, viewfinders) to see-through systems. In the mean-time, other functions can be added to the system, and if possible, why not on the display? In this section, we will also describe a new HMD see-through system with eye-tracking, embedding an OLED microdisplay with integrated sensor function.

7.4.1. *The OLED microdisplay for pico-projectors*

This development was made through the European-founded project "hypoled" [HYP 07]. Having an emissive device for a projection system is very attractive since it allows the design of a simple and compact system compared to those where light sources are required. Indeed, a compact system including the OLED microdisplay could be designed (Figure 7.22) [GRO 09].

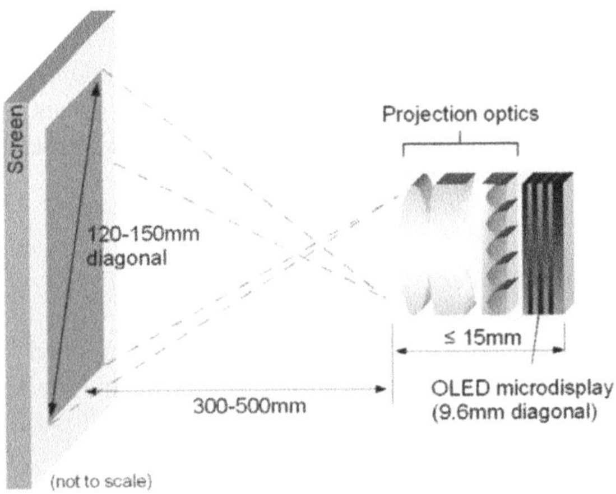

a) b)

Figure 7.22. *b) Principle system design of the ultra-compact projection system with the OLED projection engine [GRO 09] (reprinted courtesy of SID)*

To ensure the highest luminance, the choice of green monochrome OLED was made (Figure 7.23). The display had the following specification [GRO 09]: VGA (640 × 480 pixels), 12.5-μm pixel pitch, brightness: 10 000 Cd/m². Note that for achieving such high-luminance with an OLED microdisplay, compromise with lifetime has been made.

Figure 7.23. *Operating monochrome green OLED microdisplay [GRO 09](reprinted courtesy of SID)*

The system has been fabricated and evaluated [GRO 10]. The projection lens (Figure 7.24) enables an image of 90 × 120 mm (150 mm on the diagonal) with a working distance of 300 mm.

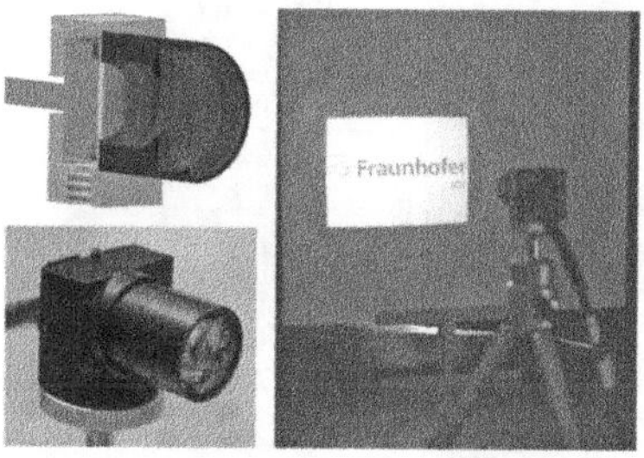

Figure 7.24. *Prototype of the monochrome one-panel pico-projection system. The adapted projection lens projects the generated image of the OLED microdisplay. The figure shows the mechanical design [GRO 10] (reprinted courtesy of SID)*

A maximal illuminance of 5.5 lx and a luminance of 0.061 lm were measured. Concerning the low luminance of the projection system, it shows that the system cannot compete with the current pico-projectors. However, the achieved image contrast of 396:1 is almost one magnitude higher than the pico-projection systems, currently on the market [GRO 10].

7.4.2. *Bi-directionnal OLED microdisplay for see-through system*

The Fraunhofer Institute for Photonic Microsystems (IPMS), in Dresden, Germany, has proposed an original approach, a full HMD see-through system, including eye-tracking, where the OLED microdisplay embeds the image sensor for the eye-tracking.

For the eye tracking, the system supplies a near-infra-red (NIR) source, which is directed into the eyes [VOG 08]. The sensor takes the image of the NIR illuminated eye. In fact, IR light provides an excellent contrast of the pupil, together with little sensitivity to ambient-light parasitic reflection.

7.4.2.1. *The OLED microdisplay with image sensor*

The idea is the integration of NIR camera photodetectors and AMOLED microdisplay into a single device, which will lead to a highly integrated, very light weight, small-sized so-called "bi-directional microdisplay". Designers consider that it is not necessary to have as much sensor pixel as display pixel.

A first generation prototype of such a device featured a display resolution of 40 × 32, intersected with a photodiode array of 20 × 16 (Figure 7.25).

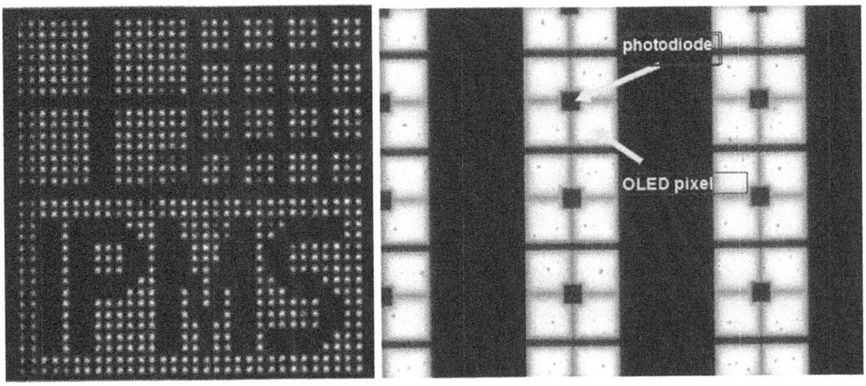

a) b)

Figure 7.25. *Bi-directional microdisplay showing pattern (left) and display/detector arrangement [VOG 08] (reprinted courtesy of SID)*

The camera-like detector function of the 20 × 16 matrix has been verified by optical test pattern (Figure 7.26).

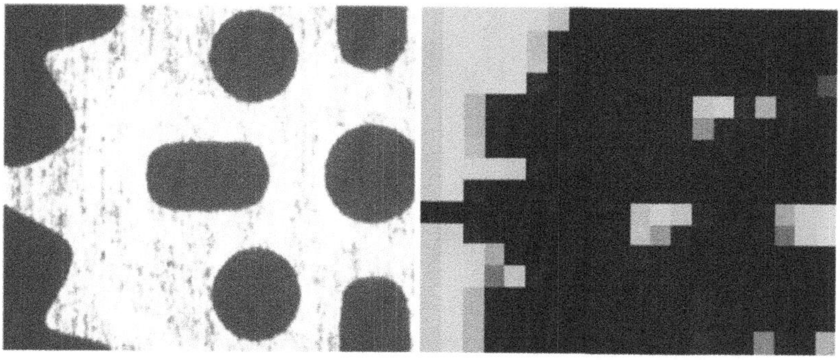

Figure 7.26. *Detector matrix function: a) test pattern and b) output signal [VOG 08] (reprinted courtesy of SID)*

After verification of functionality on this test vehicle, the next step was the design of an improved 320 × 240 (qVGA) OLED bi-directional display matrix and intersected 160 × 120 photodiode camera matrix, still having the one-to-four sensor to pixel ratio (Figure 7.27).

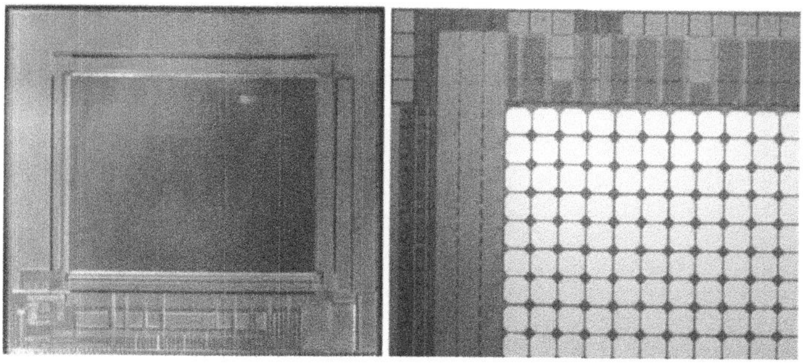

Figure 7.27. *a) Microphotographs of QVGA CMOS full device and b) part of display pixel matrix with intersected detectors [VOG 08] (reprinted courtesy of SID)*

The display diagonal is 0.61 in, with a pixel pitch of 39 μm and targeted brightness of 5000 Cd/m^2. The camera range is 600 to 900 nm.

Finally, the full-VGA version has been designed and fabricated [VOG 11]. OLED display features are 640 × 320 pixels, RGBW color structure with 16 μm pitch and camera resolution of 160 × 120 (Figure 7.28).

Figure 7.28. *Operating bi-directional VGA OLED microdisplay [p2] (reprinted courtesy of SID)*

7.4.2.2. *HMD system with eye-tracking*

The HMD system should combine three functions: near-to-eye projection (NTE), see-through functionality and eye-tracking (ET), including IR illumination of the eye. In general, see-through systems with ET use separate display and camera. In this approach, the camera pixels shrank with the OLED matrix, enabling bidirectional use of the see-through optics with an additional lens array attached to the display cover glass, realizing a compound-eye camera. The optic system is realized using a free-form prism using total internal reflection (Figure 7.29) [VOG 11].

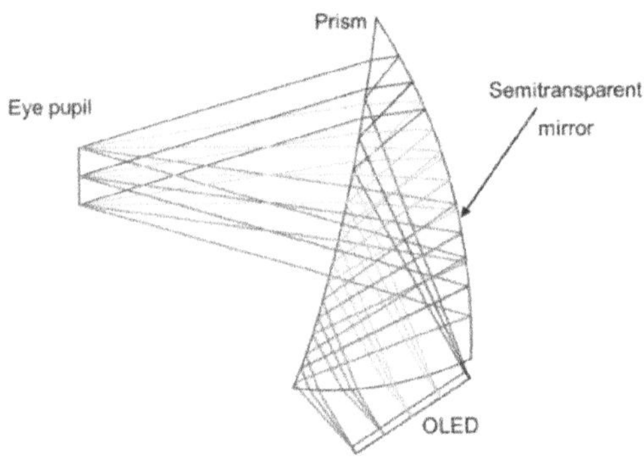

Figure 7.29. *NTE projection with free-form prism [VOG 11] (reprinted courtesy of SID)*

A second free-form correction prism is attached to the semi-transparent prism surface to enable undisturbed see-through lens. The semi-transparent mirror is realized by very thin metallization. Together with the OLED driving electronics PCB, this cage is applied to commercial sports goggles (Figure 7.30).

Figure 7.30. *Near-to-eye projection optics attached to commercial sports goggles [VOG 11] (reprinted courtesy of SID)*

Such HMD systems with eye-tracking systems, were available on demonstrator kits [COU 14].

7.4.2.3. *Other systems with bi-directional OLED microdisplays*

Bidirectional OLED microdisplays offer interesting possibilities for new applications. The compact and monolithic design of OLEDs on CMOS backplane with photo diodes enables compact system realizations, especially in the field of optical metrology. A new optical concept of an inverse confocal sensor based on bidirectional OLED microdisplays has been proposed. This system, under development, could be dedicated to applications such as the quality control measurement industry [GRO 11].

7.5. Bibliography

[CAK 06] CAKMAKCI O., ROLLAND J., "Head-worn displays: a review", *Journal of Display Technology*, vol. 2, no. 3, September 2006.

[COM 14] www.comedd.fraunhofer.de/content/dam/comedd/common/products/COMEDD/binocular-eval-kit-e.pdf, 2014.

[GRO 09] GROSSMANN C., VOGEL U., LIPPMANN U., *et al.*, "Ultra compact projection system for mobile application", *SID 2009 Digest*, pp. 1774–77, 2009.

[GRO 10] GROSSMANN C., RIEHEMANN S., NOTNI G., *et al.*, "OLED-based pico-projection system", *Journal of the SID*, vol. 18, no. 10, p. 821, 2010.

[GRO 11] GROSSMANN C., PERSKE F., ZWICK S., *et al.*, "Inverse Confocal sensor based on a Bidirectional OLED Display", *SID 2011 Digest*, pp. 66–69, 2011.

[HOS 96] HOSHI H., TANIGUCHI N., MORISHIMA H., *et al.*, "Off axis optical system consisting of aspherical surfaces without rotational symmetry", *Proceeding of SPIE*, vol. 2653, pp. 234–37, 1996.

[HYP 07] HYPOLED project (high-performance OLED-microdisplays for mobile multimedia HMD and projection applications), 2007. Available at: www.hypoled.org.

[LEV 07] LEVOLA T., "Steroscopic near to eye display using a single microdisplay", *Society for Information Display Conference Proceedings SID 07 Digest*, pp. 1158–1159, 2007.

[MEL 11] MELZER J.E., "Image source evaluation and selection of rugged near to eye display", *Society for Information Display Conference Proceedings SID 11 Digest*, pp. 132–35, 2011.

[MUK 08] MUKAWA H., AKUTSU K., MATSUMURA I., *et al.*, "A full color eyewear display using Holographic planar waveguides", *Society for Information Display Conference Proceedings SID 08 Digest*, pp. 89–92, 2008.

[PCT 06] AMITAI Y., PCT Patent, number PCT 2006 013565 A1, Lumus, 2006.

[ROL 05] ROLLAND J., HUA H., "Head-mounted display systems", *Encyclopedia of Optical Engineering*, pp. 1–14, 2005.

[SAR 13] SARAYEDDINE K., MIRZA K., "Surface structure light guide for see-through high performances true AR glasses", *International Display Workshop 2013 Conference Proceedings IDW'13*, Sapporo, Japan, pp. 127–30, December 2013.

[TAK 14] TAKEN from www.focusingscreen.com, 7 June 2014.

[UPA 85] UPATNEIKS J., US Patent no. 4 711, 51A Ann Anrbor University, 1985.

[US7 07] WEISSMAN P., HANDSCHY M., US patent no. 7206134 B2, Displaytech. Inc., 2007.

[VOG 08] VOGEL U., KREYE D., RICHTER B., *et al.*, "Bi-directional OLED microdisplay for interactive HMD", *Society for Information Display Conference Proceedings SID'08 Digest*, pp. 81–84, 2008.

[VOG 11] VOGEL U., HEROLD R., RICHTER B., *et al.*, "Interactive see-through augmented-reality smart display system", *Society for Information Display Conference Proceedings SID' 2011 Digest*, pp. 143–47, 2011.

8

OLED Microdisplays Present and Future

8.1. Present actors of OLED microdisplays

The list of organic light-emitting diode (OLED) microdisplay manufacturers proposing commercial products is given in Table 8.1, indicating the main characteristics of proposed products.

Name	Country	Founded	Main Products: reference/diagonal/ resolution	website
eMagin	US	1999	*OLED-XL WUXGA*: 0.856-in. WUXGA, 1944 × RGB × 1224 *OLED-XL SVGA*: 0.61-in. SVGA, 852 × RGB × 600	www.emagin.com/oled-microdisplays/
MicroOled	France	2007	*MDP02A*: 0.61-in. SXGA, 1300 × RGB × 1044 *MDP03*: 0.39-in. XGA, 1044 × RGB × 788	www.microoled.net
Sony	Japan	2010 (OLED micro-displays)	*ECX331A*: 0.5-in. XGA, 1024 × RGB × 768 *ECX332A*: 0.7-in. 720p, 1280 × RGB × 720	www.sony.net/Products/
Olightek	China	2008	*SXGA060*: 0.6-inch SXGA, 1280 × RGB × 1024 *SVGA097*: 0.97-inch SVGA, 804 × RGB × 604	www.olightek.com

Table 8.1. *List of companies manufacturing OLED microdisplays*

Chapter written by François TEMPLIER and Karim BOUZID.

8.2. Evolution and future developments for OLED microdisplays

8.2.1. *Introduction*

Although microdisplays are globally ready for mass production, certain technological points need to be reinforced in order to achieve a higher level of technology maturity. Also, present products need to be improved in the coming years to fulfill the requirements of future near-to-eye and viewfinder applications, as shown in Chapter 7. Among those objectives, we can list two categories: performance increase for better quality or comfort and performance increase for product viability and/or reliability. The following figures of merit can be classified in either one or other category or even both depending on the final application.

8.2.2. *Luminance and lifetime*

These two particular and very important points (luminance and lifetime) have to be treated jointly. Indeed, a certain number of technological advances made on one of these figures of merit directly concern the other, mainly via a third one: power efficiency. When improving power efficiency, luminance at a given voltage is higher. By the same process, having a better ratio of radiative recombination over total recombination also means the ratio of non-radiative, detrimental recombination is lowered. Thus, part of the degradation cause is tempered and lifetime increases. The following non-exhaustive list shows different evolutions that are or could be implemented in next generation microdisplays.

8.2.2.1. *Enhancement of stack power efficiency*

In order to increase stack power efficiency, we can first have a look at the OLED inner structure. Having a p-i-n type structure, for instance, will not only allow being independent from electrodes work functions for stack design but will also lower the turn-on voltage and thus increase power efficiency. The gain over the turn-on voltage can be explained mainly via two points.

First of all, let's suppose we consider a top-emitting OLED with a fixed microcavity dimension. When applying a voltage over the diode, a non-null electric field forms in all the layers where a voltage drop occurs. Inserting doped injection layers (in which no voltage drop happens) and conserving the minimum required thicknesses for emitting layers is a good way to increase the electric field at a given voltage over the useful regions. This is equivalent to increasing the voltage on a non-doped OLED. The second reason why p-i-n structure is beneficial to performance is linked to injection. Instead of having a thermally activated charge injection on both electrodes, the doped layers allow efficient tunneling injection even for high current densities. Moreover, not using doped injection layers implies specific engineering of electrodes work functions and limitations in the choice of materials.

Increasing power efficiency can also be achieved by a better charge confinement. The use of a hole blocking layer (HBL) and eventually an electron blocking layer (EBL) before and after the emitting layers is then advised. The probability of hole–electron recombination is higher when both species are significantly located in the same region.

Charge balance is important to achieve optimal efficiency for a given structure, but it is also crucial not to allow massive charge accumulation at critical interfaces for lifetime matter. The accumulation of either charge specie creates non-reversible oxidation-reduction reactions that will alter the chemical integrity of the structure. It has been shown that the accumulation of holes on an interface with Alq3 created secondary chemical species from Alq3 [POP 02].

Finally, recent innovative structures were proposed to broaden the recombination zone and improve carrier recombination probability within emitting layers (Figure 8.1). Such structures are mixed host based, including an emitting layer constituted of a mixture of the electron transport matrix and the hole transport matrix. The mix ratio is dependent on the transport properties of each layer for its respective main transport affinity. Duan *et al.* has, for instance, improved the L50 lifetime of a phosphorescent white OLED at 10000 cd/m^2 from 77 h up to 3000 h [DUA 11].

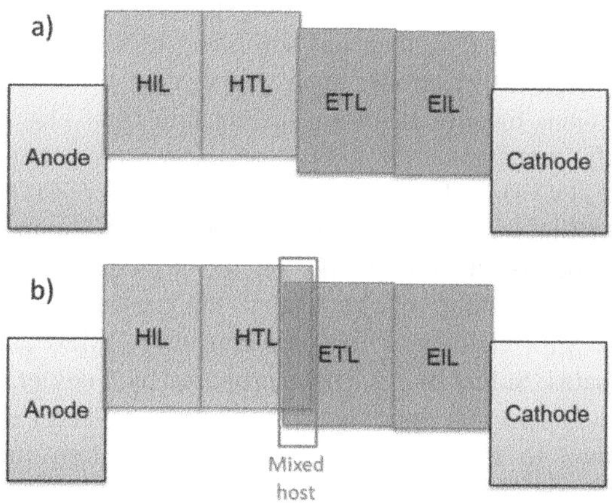

Figure 8.1. *Schematic view of a traditional OLED hetero-junction and a) a mixed host junction b) at the recombination zone. HIL stands for hole injection layer, HTL for hole transport layer, ETL for electron transport layer and EIL for electron injection layer. This figure, and all other figures in this book, is available in full color at www.iste.co.uk/templier/OLED.zip*

8.2.2.2. *Phosphorescent blue dopant*

The type of emission that occurs in the OLED has an impact over the power efficiency of the device. Phosphorescent devices are more efficient than fluorescent devices. This is true based on which population of excitons contributes in the recombination mechanisms. Phosphorescent emitters do work not only on the recombination of singulet, like fluorescent emitters do, but also on triplet excitons. The ratio between singulets and triplets is roughly 1:3. In such conditions, it is easy to perceive why the use of phosphorescent emitters entails higher power efficiency. Nevertheless, in practice, blue phosphorescent emitters tend to have the lowest lifetime among all kinds of emitters. This is the reason why most efficient OLEDs are made of a hybrid composition of emitters: fluorescent blue and phosphorescent red/green or yellow. Chemists are already working on the development of such an emitter. When a phosphorescent stable, deep blue emitter will be on the market, lifetime and power efficiency of white OLEDs will greatly benefit from it.

8.2.2.3. *Highly transparent top electrodes*

Microdisplays are, by essence, top-emission OLEDs. A microcavity effect has to be taken into account when computing the relative quantity of light generated inside the device that is actually emitted out of it. This parameter, the light outcoupling of the structure, is very dependent on the nature of the semi-transparent electrode and has a major impact on the power efficiency. Extracting more light at a given voltage means better brut performance but it can also mean the reduction of driving voltage for a given luminance requirement, thus reducing the solicitation of the device, extending its lifetime. It is considered for a thin silver electrode that the outcoupling value is only around 20%. It is then not surprising to see the emergence of new kinds of highly transparent electrodes such as dielectric-metal-dielectric structures (DMD) versus previous metallic semi-transparent thin films electrodes. These new structures, generally obtained with high n and low k dielectrics such as WO_3 (tungsten oxide) [YOO 08, HON 11] (see Figure 8.2) or ZTO (zinc tin oxide) [WIN 12] or ZnO (Zinc oxide) [LEE 12] or MoO_3 (molybdenum oxide) [ABA 13] or ZnS (Zinc Sulfide) [CHO 09], works on the principal of optical admittance optimization. The optimization of the layers' thicknesses can be addressed via optical simulation. Still, challenges like implementing DMDs as cathodes with a good injection remains. Solutions have been proposed by Yook *et al.* [YOO 08] and Hong *et al.* [HON 11]. In this case, the $WO_3/Ag/WO_3$ (WAW) electrode which is naturally predestined to hole injection rather than electron injection has been engineered with a thin aluminum or silver interlayer to work as a cathode modifying the work function seen at the interface with the organic material.

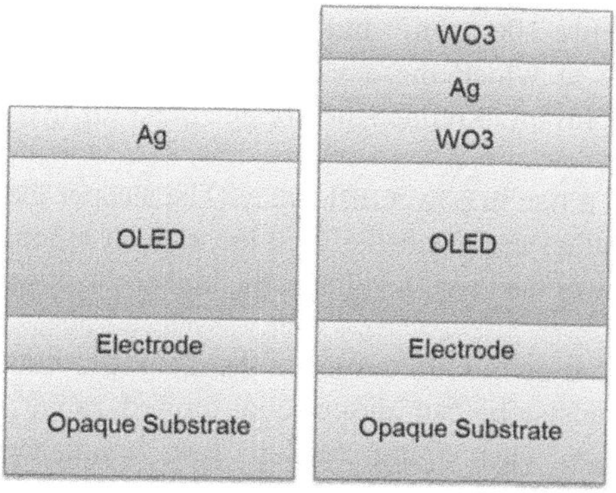

Figure 8.2. *Schematic view of a silver and a WAW electrode on top of an OLED*

8.2.2.4. *Tandem*

Last but not the least, to increase the luminance of an OLED, we can stack two units of it, creating a tandem OLED whose units are electrically connected together through a charge generation layer (CGL) [CHA 05, KID 08]. The goal here would be either to obtain more luminance for a given current density (luminance is roughly expected to increase linearly with the number of units stacked) or to drive the OLED at a lower regime for a given luminance requirement (since each unit will roughly contribute to only half the emission). The two OLED units can either be electrically connected in series or in parallel. For a given current density, units in series will need an equivalent driving voltage at least twice the one required for a single unit. For a given driving voltage, units in parallel will display an external current efficiency twice as important as the efficiency required for a single unit (each unit sees the current density entailed by the driving voltage).

The challenging aspect of this evolution is the control of the impact of the CGL over the I-V-L characteristics. In order to be usable for microdisplays, the increase in driving voltage should be as low as possible and the IV profile should not display any unusual behavior. Due to the increase of emitting layer thicknesses, for OLEDs units connected in series, the driving voltage is ideally expected to double for a given current density considering no voltage drop in the CGL [KLU 08]. At high regimes, the current density should also not be charge limited. For both reasons, the engineering of CGLs is the key to the development of efficient tandem OLEDs.

8.2.3. *Voltage delta figure of merit*

The voltage delta is a figure of merit relevant for microdisplays. It is the difference of voltage between two arbitrary luminance values, a low value and a high value, for example 100 cd/m^2 and 2000 cd/m^2, respectively. The difference between the voltages at which these values of luminance are obtained is called voltage delta. It is important to reduce the voltage delta for two independent reasons.

The first reason is a rise in power efficiency. The steeper the *I–V* curve, the faster the high-luminance values are reached. Thus, for a given luminance requirement, the steeper the *I–V*, the lower the driving voltage, the higher the power efficiency.

The second reason is due to limitations in the driving circuit. Pixels are driven by transistors. These transistors can describe a limited range of driving voltages. Unfortunately, OLEDs, when aging, tend to lose luminance at a given driving voltage. To compensate this loss in luminance, the standard solution is to increase the driving voltage. The goal for the transistor then is to be able to describe the required driving voltage without reaching the maximum voltage it can give for as long as possible. Having a steep *I–V* curve allows for a drastic increase in the current density (i.e. luminance) without having to increase the voltage by a great amount, thus allowing compensation for a longer use-time.

8.2.4. *Color coverage*

Color coverage is a figure of merit of the microdisplay giving an idea of how accurate the color representation of an image is, regarding the spectrum emitted by the pixels. The more saturated the RGB pixels are, the greater the color coverage will be. More color combinations will be possible. For white color filtered microdisplays, the color saturation is tied to the EL spectra of the emitters and the result of the intersection of the filter transmittance with the white spectrum of the OLED. Emitter EL spectra and color filter transmittance spectrums are hardly narrow. Thus, the resulting saturation of a color from a pixel can easily be polluted by neighboring emitted wavelengths. Here a way to obtain increased color coverage by adjusting the microcavity effect to the profit of each color emission is presented.

8.2.4.1. *Adjusted microcavities*

The technique is having pixels with adjusted cavity dimensions for each color. The OLED stack would intrinsically emit a white spectrum, but the engineering of the microcavity would only allow the emission of a specific part of the spectrum. Microcavity effects being more selective (spectrally speaking) than color filters, color spectrums would be more restrained in width. Color saturation for each pixel should thus increase. Furthermore, the use of color filters becoming non-critical and the loss in luminance associated to the absorbance in these layers would be retrieved. The challenge here is to be able to define different microcavity dimensions for adjacent pixels at the resolution of microdisplays. A solution has been proposed by Dodabalapur *et al.* [DOD 96] and Meerholz *et al.* [MEE 03]. Using a photo-patternable hole transport layer as described in [MEE 03] would allow such engineering as depicted in Figure 8.3.

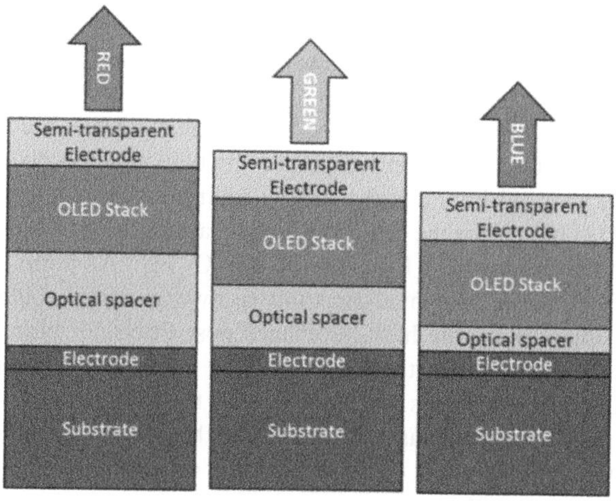

Figure 8.3. *Pixels scheme on how to obtain RGB colors with white-OLED and optical spacers*

8.2.4.2. *Color stability*

Color stability for a microdisplay is a critical point for those who aim to produce high-quality products. Yet we have to differentiate color stability versus voltage/current density and color stability during electrical aging.

8.2.4.2.1. Color stability versus voltage

In the case of a bi-emitter yellow/blue white OLED, the color point is not constant according to the driving voltage. Depending on the latter, the white emission can be slightly red-shifted at lower voltages and blue-shifted at higher voltages. This can be a struggle when trying to describe a wide range of luminance with a fixed color point.

Blom *et al.* proposed a comprehensive approach to this phenomenon in [BLO 11]. To sum up his paper, the recombination processes for blue and yellow emitters are different. Via the determination of the ideality factor for each type of emission, a discrimination of the recombination mechanism can be made. Classification as Langevin recombination or Shockley–Read–Hall (SRH) recombination is then possible. These mechanisms have different dependencies over the driving voltage. The variation of contribution of each emitter in function of driving voltage is therefore explained.

A solution can be to build up OLED stacks with emitters that emit with the same recombination processes. A complementary explanation would concern the difference in the amount of energy between the gaps of a yellow and a blue dopant. Yellow or red emitters have lower energy gaps than blue emitters, making it easier to obtain yellow electroluminescence at low voltages than blue.

The problem with voltage dependence of the color point is the following. The voltage needed to obtain the same luminance for blue and red pixels is different. (Red is obtained with red filter and the yellow contribution). The driving voltage of the white OLED is higher for the red pixel in this case because of the recombination rate dependence as shown by Blom *et al.* [BLO 11]. At high luminance, the contribution of yellow in the white spectrum is lower than the contribution at low luminance. To obtain the same high-luminance post filters for blue and red pixels and in order to compensate for the loss of yellow contribution, we would need to drive the OLED under red pixels higher. This will favor a faster degradation of red pixels versus blue pixels. Thus, in this case, having different recombination mechanisms for each emitter will entail preferential degradation of a kind of pixel.

8.2.4.2.2. Color stability over lifetime

Given a large number of images, it can be observed that pixels are roughly equivalently turned on, and thus aged. For our case of white OLED plus color filters, we can thus infer that the stress inducing degradation is homogeneous. Having different lifetimes for each emitter will be detrimental for the pixels emitting the color based on the weakest emitter. These pixels OLED stack will age faster since the degradation stress is the same for all pixels. It is then convenient to consider the microdisplay lifetime as a chain of elements. Its value is the one of the weakest elements. The goal would thus be to raise the lifetime of all emitters to the same order of magnitude in order to not have a preferential degradation of one kind of pixel.

8.2.5. *Pixel pitch size*

Lowering the pixel pitch dimension is critical to improve the screen resolution and part of the quality of the visual information it carries. For instance, for a viewfinder application, where the microdisplay is used to replace optics, the screen has to ideally be as resolved as the previously implemented system, which is to say the resolution achievable by the human eye. Therefore, the ultimate goal for this type of application in terms of pixel pitch is to reach the dimension for which the final system (additional optics included) delivers an image as resolved at the limit the human eye allows. The lower limit value for the pixel pitch is heavily dependent on the optics added on the microdisplay; thus, we would need to assess the required size for his specific application. As an example, Microoled latest 0.39" XGA microdisplays feature a subpixel pitch of 3.8×3.8 µm.

8.2.6. *Cost*

Last but not the least: it is not performance-wise but money-wise. Cost reduction of OLED microdisplays is probably as important as performance improvement, particularly at the present stage.

Increasing volumes will help cost-reduction, thanks to scale effects.

It is interesting to mention the case of other comparable *(complementary metal-oxide semiconductor)* CMOS-based, above-integrated circuit (IC) products such as CMOS imagers. Initially a high-end/expansive product, then their price progressively and continuously decreased when production boosted. Today, most CMOS imagers cost less than a dollar.

8.3. Disruptive emissive microdisplays

8.3.1. *Transparent OLED microdisplays*

Transparent OLED microdisplays have not been reported so far. However, such products might be of interest for applications such as see-through systems.

Transparent OLED in conventional small/medium size have already been presented (see Chapter 3; section 3.1.4). For such displays, the approach is quite straightforward since active matrix is fabricated with TFT process on glass, which is a transparent substrate.

It is theoretically not possible to fabricate transparent OLED microdisplays since the active matrix is a CMOS circuit, made with a non-transparent silicon wafer. However, solutions for fabricating transparent active matrices with IC technologies have already been proposed in the case of liquid crystal display (LCD), transmissive microdisplays, as mentioned in Chapter 1. This consists of transferring a SOI circuit on glass. Recently a new process for fabricating high-resolution small-pitch transparent active-matrix has been proposed [TEMP 12]. It consists of fabricating first the IC on SOI substrate, transferring it on glass, removing the backside silicon, and applying post process for the final connections including the ITO pixel electrode (Figure 8.4).

Using this process, CMOS Transparent active-matrices could be fabricated (Figure 8.5), with pixel pitch as small as 5 μm.

Figure 8.4. *Process for transparent active-matrix with small pixel-pitch from SOI wafer to transfer on glass and subsequent post-process (from [TEMP 12]. Reprinted courtesy of CEA-LETI)*

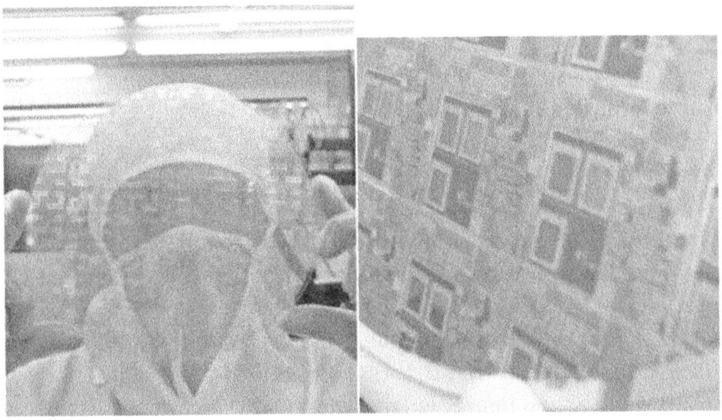

a) b)

Figure 8.5. *a) Processed glass-wafer and b) processed active-matrix made from SOI wafer transferred on glass and with post-process (from [TEMP 12]. Reprinted courtesy of CEA-LETI)*

These matrices were designed for transparent LCD microdisplays. However, to fabricate transparent OLED microdisplays, such baseline processes could therefore be used to make a transparent active-matrix, and apply OLED deposition and the rest of the process similar to existing OLED microdisplays as described in Chapter 7.

8.3.2. *Other emissive microdisplays: high-brightness GaN-based LED arrays*

8.3.2.1. *Historical background*

Group-III nitrides composed of GaN and its alloys with InN and AlN have made a breakthrough in the lighting market due to their outstanding performance and ability to emit a wide range of wavelengths in the near UV and visible spectrum.

GaN is a direct band gap semiconductor material with a 3.45 eV band gap which corresponds to 364 nm (near ultraviolet (UV)).

It was first investigated by researchers at RCA for application as a light-emitting diode (LED) [PAN 70] and later in Japan at Nagoya University [AKA 92] and Nichia Corporation by Nakamura [NAK 93].

III-Nitride LEDs have achieved dramatic advances alongside developments in indicators and solid-state lighting (SSL), and has become a viable light source for lighting applications such as automotive headlights and interior/exterior lighting. Progressively, InGaN-based white LEDs have achieved a luminous efficacy of 150 lm/W [NAR 07], which is much higher than those of other self-emissive devices including OLEDs.

Other good properties such low voltage operation, long operational lifetime (>100 000 h) and extreme conditions (high and low temperature and high humidity) operability made them good candidates for high-brightness display application.

8.3.2.2. *Application to displays*

Jiang and team at Kansas State University made extensive work toward this goal and first demonstrated microsize III–nitride LED arrays using InGaN/GaN quantum well (QW) and patterning processes [JIN 00]. Diodes had a size of 12 μm in diameter (Figure 8.6) and quantum efficiencies enhanced over the conventional broad-area LEDs, which was interpreted as the result of an enhanced current density and possibly microsize effects. First arrays of 10 × 10 diodes were also realized [JIA 01].

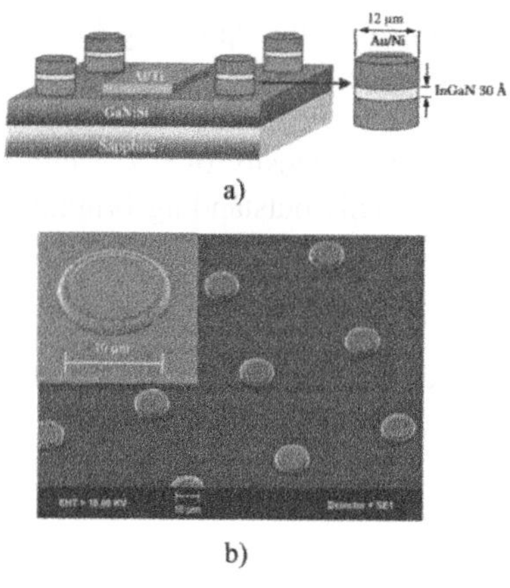

Figure 8.6. *Schematic diagram and SEM image of InGaN/GaN QW microsize LEDs (from [JIN 00])*

It was not before 2011 that actual full-resolution active-matrix III-N display could be demonstrated by a team composed of University of Texas, III-N Company and the US Army [DAY 11]. The approach was to hybridize a GaN-based LED array with an IC active matrix using the flip-chip technology via Indium bumps (Figure 8.7).

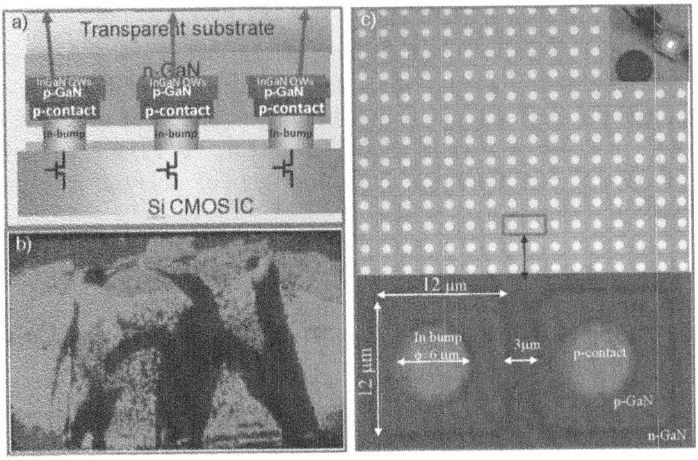

Figure 8.7. *a) Flip-chip bonding between μLED matrix array and CMOS driver IC, b) grayscale projected image from green InGaN microdisplay (pixel size 12μm and pitch 15μm) and c) zoom-in image of segment of InGaN μLED array chip showing μLED pixels and indium bumps (from [DAY 11])*

Using this process, blue (462 nm) and green (517 nm)-emitting arrays were fabricated. Green can be obtained by incorporating In in GaN, which leads to gap reduction. With the green display, each pixel of 15 μm pitch can provide brightness as high as ~4×10^6 cd/m² for a drive current of 1 μA (Figure 8.8). This current corresponds to a rather low current density of 0.7 A/cm² (1 μA)/(12 μm)²) which strongly favors device lifetime. This outstanding brightness at least two orders of magnitude higher than LCD and OLED microdisplays which provide in green ~10^4 and 10^3 cd/m², respectively.

Other groups have obtained very promising results. An important thing is that recently new companies involved in 3-5-based light-emitting arrays have been created, often being spin-offs of universities' research groups; we can mention: III-N Technologies (US), mLED (Scotland), Lumiode (US) and InfiniLED (Ireland).

Figure 8.9 shows a high-brightness, blue-emitting array. Also, small active-matrix, high output-power demonstrators have been made (Figure 8.10).

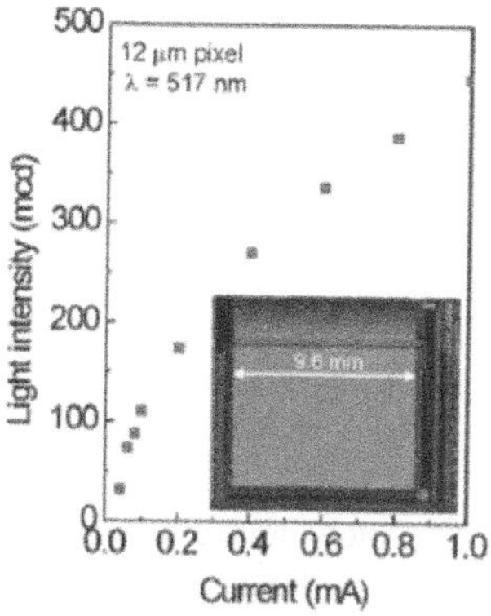

Figure 8.8. *Luminance of a single green (517 nm) µLED pixel as a function of driving current and the inset shows a micrograph of a fully assembled video graphics array (VGA) microdisplay (640 × 480 pixels). (From [DAY 11])*

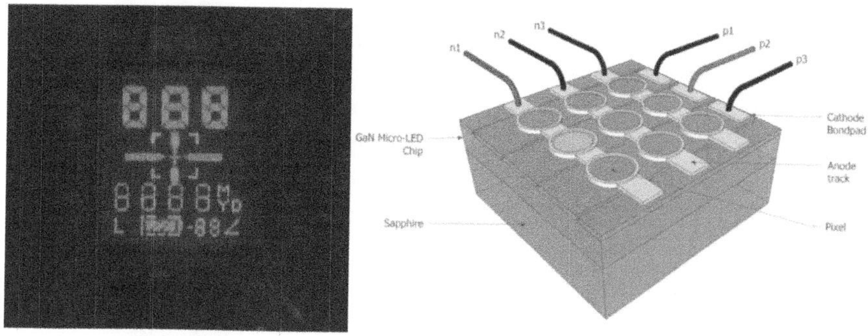

Figure 8.9. *Passive-matrix, 4096 pixels array, with 16 µm emitter on 50 µm pitch. Luminance capability: 30,000 Cd/m2 (450 nm). (Reprinted courtesy of mLED)*

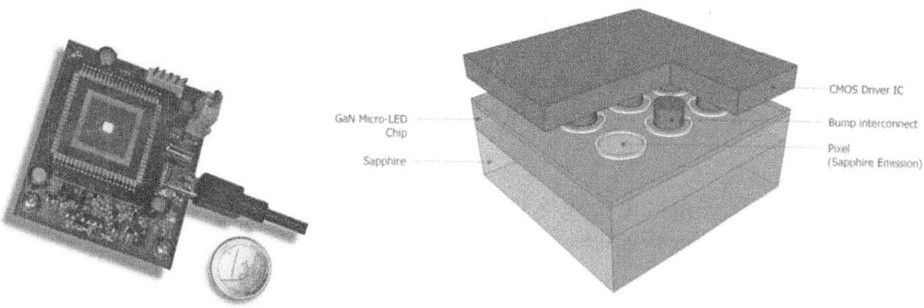

Figure 8.10. *8 × 8 micro pixelated CMOS driven microdisplay. Output optical power: 1,300 to 5,000 mW/mm^2 at 450 nm. (Reprinted courtesy of mLED)*

8.3.2.3. *Challenges for GaN-based displays*

One difficulty to fabricate such LED arrays is the poor contact resistance of the metal/*p*-type GaN interface. This is frequently a problem for III-nitride emitters due to the relatively large activation energy of Mg acceptors and low conductivity of *p*-GaN. Actually, only a small proportion (~ 1%) of Mg dopant can be electrically activated, which makes it difficult to achieve low contact resistance, which will be the way to reduce threshold current and heat generation, therefore increasing device lifetime.

But the major challenge for such new emitting devices is to obtain full-color displays. The first approach is to use direct emission. A group of researchers led by the Institute of Optics (Strathclyde University, Glasgow, Scotland) showed that by stacking a multi-QW layer emitting at 460 nm, and a multi-QW layer emitting at 600 nm, a color variation from a red (CIE: 0.54, 0.41) to a green (CIE: 0.35, 0.58) can be obtained by increasing the current density [ZHA 12] (Figure 8.11).

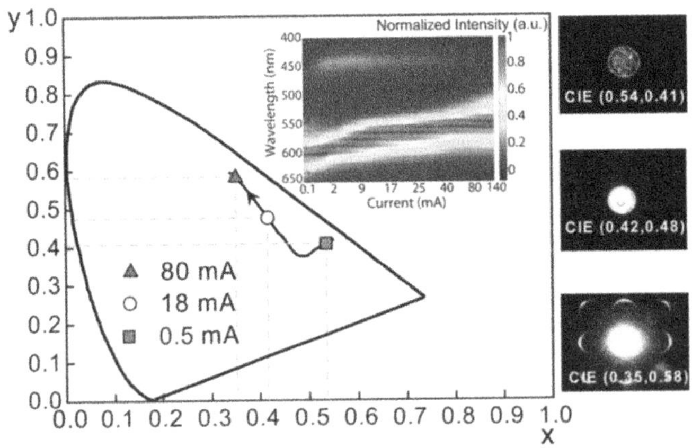

Figure 8.11. *Chromatic coordinate curve corresponding to the normalized EL spectra shown in the inset, with three specific chromatic coordinates highlighted. The three LED images are related to the highlighted chromatic coordinates. Inset: current-dependent (from 0.1 mA to 140 mA) normalized EL spectral mapping of the LED array (from [ZHA 12], reprinted with permission of Strathclyde University Glasgow)*

Despite this very encouraging result, improvements are needed before obtaining the highly saturated colors needed for display application.

The other approach is to make color down-conversion. By using appropriate photo converters, it is actually quite easy to obtain red or yellow from UV/blue/green emission. Then, the main challenge will be to structure such converters on each appropriate pixel at a very small pixel-pitch inherent to these microdisplays, namely 5 to 15 μm or less.

Such displays require hybridization techniques, which should be compatible with small pixel-pitches inherent to microdisplays. One difficulty is that so far the two substrates to be coupled (sapphire and silicon) have different thermal coefficients of expansion, which imposes a hybridization method at low/moderate temperature. For this reason, and for the question of cost, it will be very interesting for these displays to be able to use GaN structures grown on silicon instead of sapphire.

8.4. Bibliography

[ABA 13] ABACHI T., CATTIN L., LOUARN G., *et al.*, "Highly flexible, conductive and transparent MoO_3/Ag/ MoO_3 multilayer electrode for organic photovoltaic cells", *Thin Solid Films*, vol. 545, pp. 438–444, 2013.

[AKA 92] AKASAKI I., AMANO H., ITOH K., *et al.*, "GaN-based ultraviolet/blue light-emitting devices", *Institute of Physics Conference Series*, vol. 129, pp. 851–856, 1992.

[BLO 11] WETZELAER G.A.H., KUIK M., NICOLAI H.T., *et al.*, "Trap-assisted and Langevin-type recombination in organic light-emitting diodes", *Physical Review B*, vol. 83, p. 165204, 2011.

[CHA 05] CHANG C., CHEN J., HWANG S.W., *et al.*, "Highly efficient white organic electroluminescent devices based on tandem architecture", *Applied Physics Letters*, vol. 87, p. 253501, 2005.

[CHO 09] CHO H., YUN C.C., PARK J., *et al.*, "Highly flexible organic light-emitting diodes based on ZnS/Ag/WO3 multilayer transparent electrodes", *Organic Electronics*, vol. 10, pp. 1163–1169, 2009.

[DAY 11] DAY J., LI J., LIE D.Y.C., *et al.*, "III-Nitride full-scale high-resolution micro-displays", *Applied Physics Letters*, vol. 99, p. 031116, 2011.

[DOD 96] DODABALAPUR A., ROTHBERG L., JORDAN R. *et al.*, "Physics and applications of organic microcavity light-emitting diodes", *Journal of Applied Physics*, vol. 80, pp. 6954–6964, 1996.

[DUA 11] DUAN L., ZHANG D.K., *et al.*, "Controlling the recombination zone of white organic light-emitting diodes with extremely long lifetimes", *Advanced Functional Materials*, vol. 21, pp. 3540–3545, 2011.

[HON 11] HONG K., KIM K., KIM S., *et al.*, "Optical properties of wo_3/ag/wo_3 multilayer as transparent cathode in top-emitting organic light-emitting diodes", *Journal of Physical Chemistry C*, vol. 115, pp. 3453–3459, 2011.

[JIA 01] JIANG H.X., JIN S.X., LI J., *et al.*, "III-nitride blue microdisplays", *Applied Physics Letters*, vol. 78, p. 1303, 2001.

[JIN 00] JIN S.X., LI J., LI J.Z., *et al.*, "GaNmicrodisk light-emitting diodes", *Applied Physics Letters*, vol. 76, p. 631, 2000.

[KID 08] LEE T., NOH T., KIDO J., *et al.*, "High-efficiency stacked white organic light-emitting diodes", *Applied Physics Letters*, vol. 92, p. 043301, 2008.

[KLU 08] KLUBEK K., LIAO L., "Power efficiency improvement in a tandem organic light-emitting diode", *Applied Physics Letters*, vol. 92, p. 223311, 2008.

[LEE 12] LEE S., CHOI C., CHOI K., *et al.*, "Low resistive transparent and flexible ZnO/Ag/ZnO/Ag/WO_3 electrode for organic light-emitting diodes", *Organic Electronics*, vol. 13, pp. 1654–1659, 2012.

[MEE 03] MEERHOLZ K., FROHNE H., NUYKEN O., *et al.*, "Multi-colour organic light-emitting displays by solution processing", *Nature*, vol. 421, pp. 829–832, 2003.

[NAK 93] NAKAMURA S., SENOH M., IWASA N., *et al.*, "High-power InGaN single-quantum-well-structure blue and violet light-emitting diodes", *Physica B*, vol. 185, p. 453, 1993.

[NAR 07] NARUKAWA Y., SANO M., ICHIKAWA M., *et al.*, "Improvement of luminous efficiency in white light emitting diodes by reducing a forward-bias voltage", *Japanese Journal of Applied Physics*, vol. 46, p. L963, 2007.

[PAN 70] PANKOVE J., BERKEYHEISER J.E., MARUSKA H.P., *et al.*, "Luminescent properties of GaN", *Solid State Communications*, vol. 8, p. 1051, 1970.

[POP 02] POPOVIC Z, AZIZ H., "Reliability and degradation of small molecule-based organic light-emitting devices (OLEDs)", *IEEE Journal on Selected Topics in Quantum Electronics*, vol. 8, no. 2, pp. 362 – 371, March/April 2002.

[TEM 12] TEMPLIER F., ROSSINI U., SEGURA-PUCHADES J., *et al.*, "Development of 5 µm-pixel pitch active-matrix for transmissive LCD picoprojector", *The 19th International Display Workshops in conjunction with Asia Display (IDW/AD '12)*, Kyoto, Japan, 4–7 December 2012.

[WIN 12] WINKLER T., SCHMIDT H., FLUGGE H., *et al.*, "Realization of ultrathin silver layers in highly conductive and transparent zinc tin oxide/silver/zinc tin oxide multilayer electrodes deposited at room temperature for transparent organic devices", *Thin Solid Films*, vol. 520, pp. 4669–4673, 2012.

[YOO 08] YOOK K., JEON S., JOO C., *et al.*, "Transparent organic light-emitting diodes using a multilayer oxide as a low resistance transparent cathode", *Applied Physics Letters*, vol. 93, p. 013301, 2008.

[ZHA 12] ZHANG S., GONG Z., MCKENDRY J.J.D., *et al.*, "CMOS-controlled color-tunable smart display", *IEEE Photonics Journal*, vol. 4, no. 5, p. 1639, 2012.

Conclusion

After a brief overview of microdisplay technologies and applications (Introduction), we have seen that organic light-emitting diode (OLED) microdisplays, a unique monolithic emissive device, have a particular interest for near-to-eye applications, mainly for image quality, compactness and low-consumption features.

OLED microdisplays are a new practical application of these recent organic-based light-emitting devices which could be extensively developed after pioneering work which emerged in the late 80s. In Chapter 1, the basics and fundamentals of these organic-LEDS where explained, together with their main electrical and electro-optical characteristics, the different structures and their specific issues.

OLED microdisplays were, in 2001, the first commercially available active-matrix OLED displays, before their conventional-size counterparts made with TFT technologies on large area glass. However, since then, as shown in Chapter 2, small and medium size active-matrix organic light-emitting diode (AMOLED) displays have reached a much wider field of application, multibillion markets, which is expected to continue with TV application, where the competition with liquid crystal display (LCD) will be very hard.

To assess OLED display performance, it is necessary to understand the way they can be characterized: we had, in Chapter 3, an extensive review of such methods from an optical, electrical and electro-optical point of view. In the meantime, to develop new OLED displays, it is necessary to be able to predict how they will behave and perform, according to specifications, and as a function of all the OLED-stack parameters. This is covered in detail in Chapter 4.

Chapter written by François TEMPLIER.

From a product perspective, we need to know how the images are written on the display using the different addressing schemes existing (passive of active-matrix), and consequently how the driving electronics have to be designed, in particular at the pixel-level where many devices need to be integrated inside a very tiny surface. This was explained in Chapter 5, including examples of actual OLED microdisplay products.

One of the "main courses" of this book was probably Chapter 6, where we have tried to describe with the best accuracy the way OLED microdisplays are fabricated. We have seen that to make such devices, the challenge consists of successfully combining and adapting technologies from the complementary metal-oxide semiconductor (CMOS) world and from the large area display industry.

Then, how do we use and integrate these high-performance, low-consumption and compact, multicolor, display-jewels? Chapter 7 intended to answer this question by reviewing in detail the two main targeted applications: head-mounted systems and electronic view-finders. In each case, system design requirements were detailed, together with future challenges for the next-generation products. From this we could highlight the key features to improve on the OLED microdisplay side.

Chapter 8, after a brief overview of the present OLED microdisplay manufacturers, tried to list which research areas are addressed in order to reach the expected performances, namely in terms of brightness, lifetime, color performance and cost.

Today, we can summarize the situation of OLED microdisplays as follows:

– They have demonstrated unprecedented and unrivalled performance, especially in terms of:

- image quality (contrast, color, etc.);
- low power consumption;
- compactness.

– They started to find their place in at least two applications:

- head-mounted displays for professional and consumer markets;
- electronic view-finder for high-end cameras.

The short-term challenges for OLED microdisplays are the following:

– confirm and extend market presence, toward more near-to-eye applications. This should be achieved particularly where compactness is mandatory;

– diminish cost: this should come with improvement of yields, increasing throughputs/higher manufacturing volumes, etc.

The rise (not to say explosion) in interest for see-through systems is a very good opportunity for OLED microdisplays.

To achieve this, they will need to drastically boost their performances, particularly in brightness. There is no doubt that all present and future OLED microdisplays actors will do their best to be part of what could be the next display revolution.

List of Authors

Karim BOUZID
CEA-LETI
Grenoble
France

Rigo HEROLD
Westsächsische Hochschule
Zwickau
Germany

Ersun KARTAL
Germany

Philipppe LEROY
Technicolor
France

Tony MAINDRON
CEA-LETI
Grenoble
France

Christophe PRAT
CEA-LETI
Grenoble
France

Khaled SARAYEDDINE
Optinvent
France

François TEMPLIER
CEA-LETI
Grenoble
France

David VAUFREY
CEA-LETI
Grenoble
France

Index

N O, P

R, S

T

Printed and bound by CPI Group (UK) Ltd, Croydon, CR0 4YY

13/09/2026

14961082-0001